nothing

BECAUSE EVERY BOOK IS A TEST OF NEW IDEAS

A book from **NewScientist**

edited by Jeremy Webb

Nothing: *Surprising Insights Everywhere from Zero to Oblivion*

First published in the UK by Profile Books Ltd., 2013

The Experiment, LLC
220 East 23rd Street, Suite 301
New York, NY 10010-4674
www.theexperimentpublishing.com

The Experiment's books are available at special discounts when purchased in bulk for premiums and sales promotions as well as for fund-raising or educational use. For details, contact us at info@theexperimentpublishing.com.

Library of Congress Cataloging-in-Publication Data

Nothing : surprising insights everywhere from zero to oblivion / edited by Jeremy Webb.
pages cm
"First published in Great Britain in 2013 by Profile Books Ltd"--Title page verso.
Includes bibliographical references and index.
ISBN 978-1-61519-205-2 (pbk.) -- ISBN 978-1-61519-206-9 (ebook)
1. Nothing (Philosophy) 2. Science--Philosophy. I. Webb, Jeremy, 1958-
Q175.32.N68N68 2014
501--dc23

2014002254

ISBN 978-1-61519-205-2
Ebook ISBN 978-1-61519-206-9

Cover based on a design by Keenan
Cover image © Manfred_Konrad/Getty Images
Text design by Sue Lamble
Original *New Scientist* illustrations redrawn by Cherry Goddard

Manufactured in the United States of America
Distributed by Workman Publishing Company, Inc.
Distributed simultaneously in Canada by Thomas Allen & Son Ltd.

First printing March 2014
10 9 8 7 6 5 4 3 2

Contents

4 Surprises

5 Voyages of discovery

6 Conclusions

Introduction

Here's a puzzle for you: what do the big bang, a curse of death, men's nipples, antimatter traps, superconductors, penguin chicks and xenon have in common? The answer is, of course, nothing.

That is to say, I don't mean they are unrelated in any way. Quite the opposite. They are all connected by the notion of nothing—*nada, nichts, niente.*

You might think a book about nothing sounds suspiciously like an oxymoron. But fortunately there's plenty to explore, because nothing has been a topic of discussion for more than 2,000 years: indeed, the ancient Greeks had a lively disagreement about it. And such have been the changing fortunes of nothing that you can pretty much tell where you are in history just by finding out the prevailing views on nothing.

Take zero, for example, the symbol for the absence of things. Part of it came into being in Babylonia around 300 BC. The rest of it emerged 1,000 years later when the Indians fused that idea with an ancient symbol for nothingness. Another 400 years passed before it arrived in Europe where it was initially shunned as a dangerous innovation. By the 17th century it had gained acceptance, and today it is critical to the definition of every number you use.

You'll find out about all these events in the pages that follow. But there's much more besides.

The word "nothing" is applied in all manner of settings and in every case it reveals a different aspect of reality. Can something really come from nothing? Why do some animals spend all day doing nothing? What happens in our brain when we try to think about nothing? These are all questions scientists have asked and gained intriguing results.

In this way, nothing becomes a lens through which we can explore the universe around us and even what it is to be human. It reveals past attitudes and present thinking.

One example is the vacuum, the void, which is what the Greeks argued about all those centuries ago. First it didn't exist, then in the 17th century it did. During the 18th century, it was filled with a mysterious substance called luminiferous aether. That was thrown out at the start of the 20th century. But by 1930, the void had become the vacuum of quantum theory, which is about as far from nothing as you can get—it is a space packed with particles popping into and out of existence.

As this example demonstrates, nothings are usually extremes. They tend to sit at one end of a spectrum. And when scientists want to explore a phenomenon they look for an extreme version of it, because the contributory factors are often easier to spot. So if you want to measure the impact of inactivity on the body, you send your subjects to bed for a long time and order them to do absolutely nothing. The results of that particular experiment changed medical practice overnight.

Another extreme is absolute zero, the coldest cold that can exist, where the thermal jiggling of atoms all but disappears. Our journey toward absolute zero has been a tortuous one, filled with misconceptions and blind alleys. Yet the human impulse to explore eventually revealed a world of bizarre behaviors that we could never have predicted.

Nothings can be difficult to attain: we haven't reached absolute zero and most likely never will. Nothings can also be messy: what is described as the vacuum of space turns out to be not one, but many. And nothings can be powerful: sick people can get better after talking with a doctor even though nothing material passes between them. This effect, which is perplexing some of the best brains in medical science, has an equally powerful evil twin.

These are just a few ways in which nothing can reveal glimpses of our universe. It would have been relatively easy to corral these stories into chapters themed along conventional lines—cosmology, mathematics and so on. But in *New Scientist*, where most of these essays originated, we have found that variety is highly prized and it is always wise for every issue of the magazine to offer something for everyone.

In that spirit, I have instead created chapters around topics such as beginnings, mysteries and surprises. So if physics is not your bag, it won't be long before you reach something more to your taste. I hope to intrigue you with the sheer breadth of the ways in which nothing has influenced our thinking.

Themes such as the birth and death of the universe, the vacuum, the power of nothing, zero and absolute zero run through the chapters. For those who wish to read all the essays on a specific theme, there's a signpost at the end of each essay pointing you to the next one in the chain.

One use of the word nothing implies a lack of value: if something is insignificant, people say "it's nothing." That meaning clearly comes from a time before we realized quite how valuable nothing is. I hope I can convince you it is a concept rich in meaning and implication.

Jeremy Webb

1

Beginnings

"Astronomy leads us to a unique event, a universe which was created out of nothing," said Arno Penzias, the American physicist and Nobel laureate. He was talking about the mother of all beginnings, the big bang. It's the obvious place for us to start. To add some variety, we'll bounce you to ancient Babylon and then to the most modern of brain-scanning laboratories. You'll find out about the birth of a symbol that you almost certainly take for granted and discover that your head is home to an organ you've probably never heard of. Along the way, we'll look at the fruits of an infant scientific field—the mind's power to heal the body.

The big bang

Our universe began in an explosion of sorts, what's called the big bang. The $64,000 question is how the cosmos emerged out of nothing. But before we tackle that, we need to understand what the big bang entailed. Here's Marcus Chown.

In the beginning was nothing. Then the universe was born in a searing hot fireball called the big bang. But what was the big bang? Where did it happen? And how have astronomers come to believe such a ridiculous thing?

About 13.82 billion years ago, the universe that we inhabit erupted, literally, out of nothing. It exploded in a titanic fireball called the big bang. Everything—all matter, energy, even space and time—came into being at that instant.

In the earliest moments of the big bang, the stuff of the universe occupied an extraordinarily small volume and was unimaginably hot. It was a seething cauldron of electromagnetic radiation mixed with microscopic particles of matter unlike any found in today's universe. As the fireball expanded, it cooled, and more and more structure began to "freeze out."

Step by step, the fundamental particles we know today, the building blocks of all ordinary matter, acquired their present identities. The particles condensed into atoms and galaxies began to grow, then fragment into stars such as our sun. About 4.55 billion years ago, Earth formed. The rest, as they say, is history.

It is an extraordinarily grand picture of creation. Yet astronomers and physicists, armed with a growing mass of evidence to back their theories, are so confident of the scenario that they believe they can work out the detailed conditions in the early universe as it evolved, instant by instant.

That's not to say we can go back to the moment of creation. The best that physics can do is to attempt to describe what was happening when the universe was

Looking backward in time

Physicists can run the expansion of the universe backward. In this way, they can watch it get hotter as it gets smaller, just as the air in a bicycle pump heats up as it is compressed. But theory proposes that, at the big bang itself, the temperature was infinite. And infinities warn physicists that theories are flawed.

At the moment, the theories which take us furthest back in time are the Grand Unified Theories. These GUTs are an attempt to show that three of the basic forces that govern the behavior of all matter—the strong and weak nuclear forces and the electromagnetic force—are no more than facets of a single "superforce."

Each force of nature arises from the exchange of a different "messenger" particle, or boson. The messenger transmits a force between two particles, just as a tennis ball transmits to a player the force of an opponent's shot. At high enough temperatures—such as those when the universe was 10^{-35} seconds old—physicists believe the electromagnetic and strong and weak nuclear forces were identical, and mediated by a messenger dubbed the X-boson.

Physicists want to show that gravity, too, is a facet of the superforce. They suspect that gravity split apart from the other three forces at about 10^{-43} seconds after the big bang. But before they can "unify" the four forces, they must describe gravity using quantum theory, which is hugely successful for describing the other forces. To say that physicists are finding this difficult is an understatement.

When they have their unified theory, physicists believe that they will be able to probe right back to the moment of creation and explain how the universe popped suddenly into existence from nothing 13.82 billion years ago.

already about 10^{-35} seconds old—a length of time that can also be written as a decimal point followed by 34 zeroes and a 1.

This is an exceedingly small interval of time, but you would be wrong if you thought it was so close to the moment of creation as to make no difference. Although the structure of the universe no longer changes much in even a million years, when the universe was young, things changed much more rapidly.

For example, physicists think that as many important events happened between the end of the first tenth of a second and the end of the first second as in the interval from the first hundredth of a second to the first tenth of a second, and so on, logarithmically, back to the very beginning. As they run the history of the universe backward, like a movie in reverse, space is filled with ever more frenzied activity.

This is because the early universe was dominated by electromagnetic radiation—in the form of little packets of energy called photons—and the higher the temperature, the more energetic the photons. Now, high-energy photons can change into particles of matter because one form of energy can be converted into another, and, as Einstein revealed, mass (m) is simply a form of energy (E), hence his famous equation $E=mc^2$, where c is the speed of light.

What Einstein's equation says is that particles of a particular mass, m, can be created if the packets of radiation, the photons, have an energy of at least mc^2. Put another way, there is a temperature above which the photons are energetic enough to produce a particle of mass, m, and below which they cannot create that particle.

If we look far enough back, we come to a time when the temperature was so high, and the photons so energetic, that colliding photons could produce particles out of radiant energy. What those particles were before the universe was 10^{-35} seconds old, we do not know. All we can say is that they were very much more massive than the particles we are familiar with today, such as the electron and top quark.

As time progressed and temperature fell, so the mix of particles in the universe changed to a soup of less and less massive particles. Each particle was "king for a day," or at least for a split second. For the reverse process was also going on—matter was being converted back to radiant energy as particles collided to produce photons.

What do physicists think the universe was like a mere 10^{-35} seconds after the big bang?

Well, the volume of space that was destined to become the "observable universe," which today is 84 billion light years across, was contained in a volume roughly the size of a pea. And the temperature of this superdense material was an unimaginable 10^{28} °C.

At this temperature, physicists predict, colliding photons had just the right amount of energy to produce a particle called the X-boson that was a million billion times more massive than the proton. No one has yet observed an X-boson, because to do so we would have to recreate, in an Earth-bound laboratory, the extreme conditions that existed just 10^{-35} seconds after the big bang.

How far back can physicists probe in their laboratories?

The answer is to a time when the universe was about one-trillionth (10^{-12}) of a second old. By then, it had cooled down to about 100 million billion degrees—still 10 billion times hotter than the center of the sun. In 2012, physicists at CERN, the European center for particle physics in Geneva, recreated these conditions in the giant particle accelerator called the Large Hadron Collider. They conjured into being a particle that resembles the Higgs boson, a particle that vanished from the universe a trillionth of a second after the big bang.

The gulf between 10^{-35} seconds and a trillionth of a second is gigantic. We know that for most of this period, matter was squeezed together more tightly than the most compressed matter we know of—that inside the nuclei of atoms. And, as the temperature fell, so the energy level of photons declined, creating particles of lower and lower masses.

At some point, the hypothetical building blocks of the neutron and proton—known as quarks—came into being. And by the time the universe was about one-hundredth of a second old, it had cooled sufficiently to be dominated by particles that are familiar to us today: photons, electrons, positrons and neutrinos. Neutrons and protons were around, but there weren't many of them. In fact, they were a very small contaminant in the universe.

About one second into the life of the universe, the temperature had fallen to about 10 billion °C, and photons had too little energy to produce particles easily. Electrons and their positively charged "antimatter" opposites, called positrons, were colliding and annihilating each other to create photons. However, because of a slight and, to this day, mysterious lopsidedness in the laws of physics, there were roughly 10 billion + 1 electrons for every 10 billion positrons. So, after an orgy of annihilation, the universe was left with a surplus of matter, and with about 10 billion photons for every electron, a ratio that persists today.

The next important stage in the history of the universe was at about one minute.

The temperature had dropped to a mere 1 billion °C—the temperature in the hearts of the hottest stars. Now the particles were moving more slowly. In the case of protons and neutrons, it meant that they stayed close to each other long enough for the strong nuclear forces, which bind them together in the nuclei of atoms, to have a chance to take hold. In particular, two protons and two neutrons could combine to form nuclei of helium.

Solitary neutrons decay into protons in about 15 minutes, so any neutrons left over after helium formed became protons. According to physicists' calculations, roughly ten protons were left over for every helium nucleus that formed. And these became the nuclei of hydrogen atoms, which consist of a single proton.

This is one of the strongest pieces of evidence that the big bang really did happen. For much, much later, when the temperature had cooled considerably, the hydrogen and helium nuclei picked up electrons to become stable atoms. Today, when astronomers measure the abundance of elements in the universe—in stars, galaxies and interstellar space—they still find roughly one helium atom for every ten hydrogens.

The point at which it was cool enough for electrons to combine with protons to make the first atoms was about 380,000 years after the big bang. The universe was now cooling very much more slowly than in its early moments, and the temperature had reached a modest 3,000 °C. This also marked another significant event in the early history of the universe.

Until the electrons had combined with the hydrogen and helium nuclei, photons could not travel far in a straight line without running into an electron. Free electrons are very good at scattering, or redirecting, photons. As a consequence, every photon had to zigzag its way across the universe. This had the effect of making the universe opaque. If this happened today and light from the stars zigzagged its way across space to your eyes, rather than flying in straight lines, you would see only a dim milky glow from the whole sky rather than myriad stars.

We can still detect photons from this period. They have been flying freely through the universe for billions of years, and astronomers observe them as what's called the cosmic microwave background. Whereas these photons started their journey when the temperature was 3,000 °C, the universe has expanded about 1100 times while they have been in flight. This has decreased their energy by this factor, so that we now record the signals as just 2.725 degrees above absolute zero.

The temperature dropping to about 3,000 °C also signalled another event—the point at which the energy levels of the radiation, or photons, in the universe fell below that of the matter. From then on, the universe was dominated by matter and by the force of gravity acting on that matter.

The building of elements, which had begun when the universe was about one minute old, had stopped by the time it had been in existence for ten minutes, and the protons and neutrons had formed the nuclei of hydrogen and helium.

For elements such as carbon and oxygen to form, hotter and denser conditions were needed, but the universe was getting colder and more rarefied all the while. The heavy elements in the planets and in your body were created, billions of years later, in the nuclear furnaces of stars.

Instead, as the universe continued to expand, gravity caused clumps of matter to accumulate in large islands. Those islands were to become the galaxies. The galaxies continued their headlong rush into the void, fragmenting into smaller clumps which became individual stars, producing heat and light by nuclear reactions deep in their cores. At one point, about 9 billion years after the big bang, a yellow star was born toward the outer edge of a great spiral whirlpool of stars called the Milky Way. The star was our sun.

How do we know there was a big bang?

Our modern picture of the universe is due in large part to an American astronomer, Edwin Hubble. In 1923, he showed that the Milky Way, the great island of stars to which our sun belongs, was just one galaxy among thousands of millions of others scattered throughout space.

Hubble also found that the wavelength of the light from most of the galaxies is "red shifted." Astronomers initially interpreted this as a Doppler effect, familiar to anyone who has noticed how the pitch of a police siren drops as it passes by. The siren becomes deeper because the wavelength of the sound is stretched out. Similarly with light, the wavelength of light from a galaxy which is moving away from us is stretched out to a longer, or redder, wavelength.

Hubble discovered that most galaxies are receding from the Milky Way. In other words, the universe is expanding. And the farther away a galaxy is, the faster it is receding.

One conclusion is inescapable: the universe must have been smaller in the past. There must have been a moment when the universe started expanding: the moment of its birth. By imagining the expansion running backward, astronomers deduce that the universe came into existence about 13.82 billion years ago.

This idea of a big bang means that the red shifts of galaxies are not really Doppler shifts. They arise because in the time that light from distant galaxies has been traveling across space to Earth, the universe has grown, stretching the wavelength of light.

The picture of a universe that is expanding need not have been a surprise to anyone. If Albert Einstein had only had faith in his equations, he could have predicted it in 1915 with his theory of gravity, known as the general theory of relativity. But Einstein, like Newton before him, hung on to the idea that the universe was static—unchanging, without beginning or end. He can be forgiven because, at the time, he did not even know about the existence of galaxies.

The vision of a static universe also appealed strongly to astronomers. In 1948, Hermann Bondi, Thomas Gold and Fred Hoyle proposed the steady-state theory of the universe. The universe was expanding, they said, but perhaps it was unchanging in time.

Their theory said that space is expanding at a constant rate but, at the same time, matter is created continuously throughout the universe. This matter is just enough to compensate for the expansion and keep the density of the universe constant. Where this matter would come from, nobody could say. But neither could the proponents of the big bang.

The steady-state theory held its own as the principal challenger to the big bang theory for two decades. Then, in the 1960s, two astronomical discoveries dealt it a fatal blow.

The first discovery came from Martin Ryle and his colleagues at the University of Cambridge. They were studying radio galaxies—enormously powerful sources of radio waves. In the early 1960s, the Cambridge astronomers found that there were many more radio galaxies at large distances than nearby.

The radio waves from these distant objects have taken billions of years to reach us. Ryle and his colleagues, therefore, were observing our universe as it was in an earlier time. The excess of radio galaxies at great distances had to mean that conditions in the remote past were different from those today. A universe which changes with time ran counter to the steady-state theory.

Then in 1965, Arno Penzias and Robert Wilson, two scientists at the Bell Telephone Labs in Holmdel, New Jersey, detected an odd signal with a radio horn they had inherited from engineers working on Echo 1 and Telstar, the first communication satellites.

The signal did not come from Earth or the sun. It seemed to come from all over the sky, and it was equivalent to the energy emitted by a body at about 3 degrees above absolute zero (–270 °C).

There could be no doubt. Penzias and Wilson had discovered the "afterglow" of the big bang fireball—the cosmic microwave background. For their proof of the big bang, they shared the 1978 Nobel prize in physics.

For more cosmology, go to "The day time began" on page 44.

Secret life of the brain

People have been rummaging around inside the human body for millennia, so to find a new organ in the 21st century is an extraordinary achievement. That's effectively what two researchers have done. Their discovery came from asking a simple question: what happens when the brain is resting—when it's doing nothing? Douglas Fox takes up the story.

In 1953, a physician named Louis Sokoloff laid a 20-year-old college student onto a gurney, attached electrodes to his scalp and inserted a syringe into his jugular vein.

For 60 minutes the volunteer lay there and solved arithmetic problems. All the while, Sokoloff monitored his brainwaves and checked the levels of oxygen and carbon dioxide in his blood.

Sokoloff, a researcher at the University of Pennsylvania in Philadelphia, was trying to find out how much energy the brain consumes during vigorous thought. He expected his volunteer's brain to guzzle more oxygen as it crunched the problems, but what he saw surprised him: his subject's brain consumed no more oxygen while doing arithmetic than it did while he was resting with his eyes closed.

People have long envisaged the brain as being like a computer on standby, lying dormant until called upon to do a task, such as solving a sudoku, reading a newspaper or looking for a face in a crowd. Sokoloff's experiment provided the first glimpse of a different truth: that the brain enjoys a rich private life. This amazing organ, which accounts for only 2 percent of our body mass but devours 20 percent of the calories we eat and drink, fritters away much of that energy doing, as far as we can tell, absolutely nothing.

"There is a huge amount of activity in the [resting] brain that has been largely unaccounted for," says Marcus Raichle, a neuroscientist at Washington University in St Louis. "The brain is a very expensive organ, but nobody had asked deeply what this cost is all about."

Raichle and a handful of others are finally tackling this fundamental question—what exactly is the idling brain up to, anyway? Their work has led to the discovery of a major system within the brain, an organ within an organ, that hid for decades right before our eyes. Some call it the neural dynamo of daydreaming. Others assign it a more mysterious role, possibly selecting memories and knitting them seamlessly into a personal narrative. Whatever it does, it fires up whenever the brain is otherwise unoccupied and burns white hot, guzzling more oxygen, gram for gram, than your beating heart.

"It's a very important thing," says Giulio Tononi, a neuroscientist at the University of Wisconsin-Madison. "It's not very frequent that a new functional system is identified in the brain—in fact it hasn't happened for I don't know how many years. It's like finding a new continent."

The discovery was slow in coming. Sokoloff's experiment in 1953 drew little attention. It wasn't until the 1980s that it started to dawn on researchers that the brain may be doing important things while apparently stuck in neutral.

In the 1980s a novel brain-scanning technique called PET (for positron emission topography) was all the rage. By injecting radioactive glucose and measuring where it accumulated, researchers were able to eavesdrop on the brain's inner workings. In a typical experiment they would scan a volunteer lying down with their eyes closed and again while doing a mentally demanding task, then subtract one scan from the other to find the brain areas that lit up.

Raichle was using PET to find brain areas associated with words when he noticed something odd: some brain areas seemed to go at full tilt during rest, but quietened down as soon as the person started an exercise. Most people shrugged off these oddities as random noise. But in 1997, Raichle's colleague Gordon Shulman found otherwise.

Shulman sifted through a stack of brain scans from 134 people. Regardless of the task performed, whether it involved reading or watching shapes on a screen, the same constellation of brain areas always dimmed as soon as the subject started concentrating. "I was surprised by the level of consistency," says Shulman. Suddenly it looked a lot less like random noise. "There was this neural network that had not previously been described."

Raichle and Shulman published a paper in 2001 suggesting that they had stumbled onto a previously unrecognized "default mode"—a sort of internal game of solitaire which the brain turns to when unoccupied and sets aside when called on to do something else. This brain activity occurred largely in a cluster of regions arching through the midline of the brain, from front to back, which Raichle and Shulman dubbed the default network.[1]

The brain areas in the network were known and previously studied by researchers. What they hadn't known before was that they chattered non-stop to one another

The brain in neutral

When you switch off, the default network bursts into action

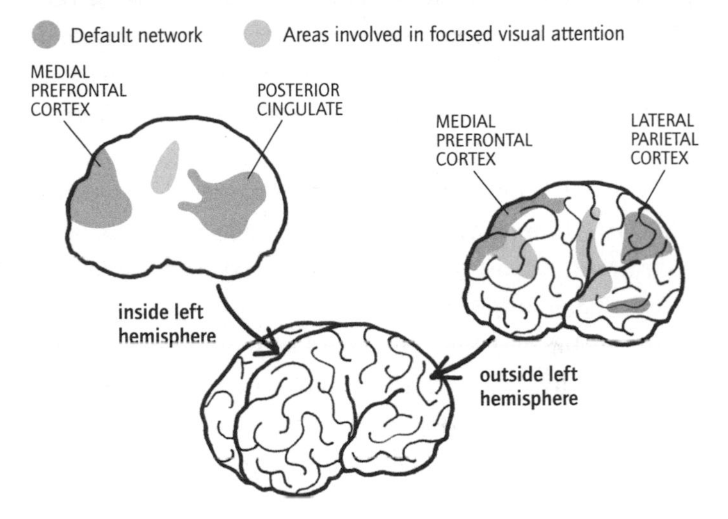

when the person was unoccupied, but quieted down as soon as a task requiring focused attention came along. Measurements of metabolic activity showed that some parts of this network devoured 30 percent more energy, gram for gram, than nearly any other area of the brain.

All of this poses the question—what exactly is the brain up to when we are not doing anything? When Raichle and Shulman outlined the default network, they saw clues to its purpose based on what was already known about the brain areas concerned.

One of the core components is the medial prefrontal cortex (see figure above), which is known to evaluate things from a highly self-centered perspective of whether they're likely to be good, bad or indifferent. Parts of this region also light up when people are asked to study lists of adjectives and choose ones that apply to themselves but not to, say, Madonna. People who suffer damage to their medial prefrontal cortex become listless and uncommunicative. One woman who recovered from a stroke in that area recalled inhabiting an empty mind, devoid of the wandering, stream-of-consciousness thoughts that most of us take for granted.

Parts of the default network also have strong connections to the hippocampus, which records and recalls autobiographical memories such as yesterday's breakfast or your first day of kindergarten.

To Raichle and his colleague Debra Gusnard, this all pointed to one thing: daydreaming. Through the hippocampus, the default network could tap into memories—the raw material of daydreams. The medial prefrontal cortex could then evaluate those memories from an introspective viewpoint. Raichle and Gusnard speculated that the default network might provide the brain with an "inner rehearsal" for considering future actions and choices.

Randy Buckner, a former colleague of Raichle's, now at Harvard, agrees. To him the evidence paints a picture of a brain system involved in the quintessential acts of daydreaming: mulling over past experiences and speculating about the future. "We're very good at imagining possible worlds and thinking about them," says Buckner. "This may be the brain network that helps us to do that."

There is now direct evidence to support this idea. In 2007, Malia Mason, now at Columbia Business School, New York City, reported that the activity of the default network correlates with daydreaming. Using the brain-imaging technique functional magnetic resonance imaging (fMRI), Mason found that people reported daydreaming when their default network was active, but not when it dimmed down. Volunteers with more active default networks reported more wandering thoughts overall.[2]

Daydreaming may sound like a mental luxury, but its purpose is deadly serious: Buckner and his Harvard colleague Daniel Gilbert see it as the ultimate tool for incorporating lessons learned in the past into our plans for the future. So important is this exercise, it seems, that the brain engages in it whenever possible, breaking off only when it has to divert its limited supply of blood, oxygen and glucose to a more urgent task.

But people now suspect that the default network does more than just daydream. It started in 2003 when Michael Greicius of Stanford University in California studied the default network in a new way. He got his subjects to lie quietly in an fMRI scanner and simply watched their brains in action. This led him to find what are called resting state fluctuations in the default network—slow waves of neural activity that ripple through in a coordinated fashion, linking its constellation of brain areas into a coherent unit. The waves lasted 10 to 20 seconds from crest to crest, up to 100 times slower than typical electroencephalograph (EEG) brain waves recorded by electrodes on the scalp.

Until then scientists had studied the default network in the old-fashioned way, subtracting resting scans from task scans to measure changes in brain activity. But Greicius's work showed that you could eavesdrop on the network by simply scanning people as they lay around doing nothing. This allowed scientists to study the network in people who weren't even conscious, revealing something unexpected.

Raichle reported in 2007 that the network's resting waves continued in heavily anesthetized monkeys as though they were awake.[3] More recently, Greicius reported a similar phenomenon in sedated humans, and other researchers have found the default network active and synchronized in early sleep.[4]

This derailed the assumption that the default network is all about daydreaming. "I was surprised," admits Greicius. "I've had to revamp my understanding of what we're looking at."

Given that the default network is active in early sleep it's tempting to link it with real dreaming, but Raichle suspects its nocturnal activity has another purpose—sorting and preserving memories. Each day we soak up a mountain of short-term memories, but only a few are actually worth adding to the personal narrative that guides our lives.

Raichle now believes that the default network is involved, selectively storing and updating memories based on their importance from a personal perspective—whether they're good, threatening, emotionally painful, and so on. To prevent a backlog of unstored memories building up, the network returns to its duties whenever it can.

In support of this idea, Raichle points out that the default network constantly chatters with the hippocampus. It also devours huge amounts of glucose, way out of proportion to the amount of oxygen it uses. Raichle believes that rather than burning this extra glucose for energy, it uses it as a raw material for making the amino acids and neurotransmitters it needs to build and maintain synapses, the very stuff of memory. "It's in those connections where most of the cost of running the brain is," says Raichle.

With such a central role, it shouldn't be surprising that the default network is implicated in some familiar brain diseases. In 2004, Buckner saw a presentation by William Klunk of the University of Pittsburgh School of Medicine. Klunk presented 3D maps showing harmful protein clumps in the brains of people with Alzheimer's disease. Until then scientists had only looked at these clumps in one brain location at a time, by dissecting the brains of deceased patients. So when Klunk projected his whole-brain map on the screen, it was the first time many people had seen the complete picture. "It was quite surprising," says Buckner. "It looked just like the default network."

Raichle, Greicius and Buckner have since found that the default network's pattern of activity is disrupted in patients with Alzheimer's. They have also begun to monitor default network activity in people with mild memory problems to see if they can learn to predict who will go on to develop Alzheimer's. Half of people with memory problems go on to develop the disease, but which half? "Can we use what we've learned to provide insight into who's at risk for Alzheimer's?" asks Buckner.

The default network also turns out to be disrupted in other maladies, including depression, attention deficit hyperactivity disorder (ADHD), autism and schizophrenia. It plays a mysterious role too in victims of brain injury or stroke who hover in the grey netherworld between consciousness and brain death known as a minimally conscious or vegetative state. Steven Laureys, a neurologist at the University of Liège in Belgium, has used fMRI to look at patterns of activity in the default networks of people in this state. "You can really see how this network breaks down as coma deepens," he says. He is now looking for a link between default network activity and whether patients will regain consciousness after, say, 12 months. "We're hoping to show that it will have prognostic value," he says.

All of this has been a long time coming since Sokoloff's surprising observation more than half a century ago. Watching the brain at rest, rather than constantly prodding it to do tricks, is now revealing the rich inner world of our private moments. So the next time you're mooching around doing nothing much, take a moment to remind yourself that your brain is still beavering away—if you can tear yourself away from your daydreams, that is.

The meditating mind

When Zen Buddhists meditate, they may be deliberately switching off their default network. The goal of Zen meditation is to clear the mind of wandering, stream-of-consciousness thoughts by focusing attention on posture and breathing.

Giuseppe Pagnoni, a neuroscientist at the Emory University School of Medicine in Atlanta, Georgia, wondered whether this meant they had learned to suppress the activity of their default network. He recruited a group of volunteers trained in Zen meditation and put them in an fMRI scanner. He presented them with random strings of letters and asked them to determine whether each was an English-language word or just gibberish. Each time a subject saw a real word, their default network would light up for a few seconds–evidence of meandering thoughts triggered by the word, such as apple, leading to apple pie, leading to cinnamon. Zen meditators performed just as well as non-meditators on word recognition, but they were much quicker to rein in their daydreaming engines afterward, doing so within about 10 seconds, versus 15 seconds for non-meditators.[5]

To read about things that do nothing, go to "Wastes of space?" on page 65.

From zero to hero

These days we take zero for granted as a symbol for an absence of things. We think of it as just another number, like 1, 2 or 3. But it's not. Europe was late to accept this invention, and for a while it was even banned. Richard Webb delves into the enigma that is zero.

I used to have seven goats. I bartered three for corn; I gave one to each of my three daughters as dowry; one was stolen. How many goats do I have now?

This is not a trick question. Oddly, though, for much of human history we have not had the mathematical wherewithal to supply an answer.

There is evidence of counting that stretches back five millennia in Egypt, Mesopotamia and Persia. Yet even by the most generous definition, a mathematical conception of nothing—a zero—has existed for less than half that time. Even then, the civilizations that discovered it missed its point entirely. In Europe, indifference, myopia and fear stunted zero's development for centuries. What is it about zero that stopped it becoming a hero?

This is a tangled story of two zeros: zero as a symbol to represent nothing, and zero as a number that can be used in calculations and has its own mathematical properties. It is natural to think the two are the same. History teaches us something different.

Zero the symbol was in fact the first of the two to pop up by a long chalk. This is the sort of character familiar from a number such as the year 2012 in our calendar. Here it acts as a placeholder in our "positional" numerical notation, whose crucial feature is that a digit's value depends on where it is in a number. Take 2012, for example: a "2" crops up twice, once to mean 2 and once to mean 2,000. That's because our positional system uses "base" 10—so a move of one place to the left in a number means a digit's worth increases by a further power of 10.

It is through such machinations that the string of digits "2012" comes to have the properties of a number with the value equal to $2 \times 10^3 + 0 \times 10^2 + 1 \times 10^1 + 2$. Zero's role is pivotal: were it not for its unambiguous presence, we might easily mistake 2,012 for 212, or perhaps 20,012, and our calculations could be out by hundreds or thousands.

The first positional number system was used to calculate the passage of the seasons and the years in Babylonia, modern-day Iraq, from around 1800 BC onward. Its base was not 10, but 60. It didn't have a symbol for every whole number up to the base, unlike the "dynamic" system of digits running from 1 to 9 that is the bread-and-butter of our base-10 system. Instead it had just two symbols, for

1 and 10, which were clumped together in groups with a maximum headcount of 59. For example, 2,012 equates to $33 \times 60^1 + 32$, and so it would have been represented by two adjacent groups of symbols: one clump of three 10s and three ones; and a second clump of three 10s and two ones.

This particular number has nothing missing. Quite generally, though, for the first 15 centuries or so of the Babylonian positional numbering system the absence of any power of 60 in the transcription of any number was marked not by a symbol, but (if you were lucky) just by a gap. What changed around 300 BC we don't know; perhaps one egregious confusion of positions too many. But it seems to have been at around this time that a third symbol, a curious confection of two left-slanting arrows (see timeline on page 31), started to fill missing places in the stargazers' calculations.

This was the world's first zero. Some seven centuries later, on the other side of the world, it was invented a second time. Mayan priest-astronomers in central America began to use a snail-shell-like symbol to fill gaps in the (almost) base-20 positional "long-count" system they used to calculate their calendar.

Zero as a placeholder was clearly a useful concept, then. It is a frustration entirely typical of zero's vexed history, though, that neither the Babylonians nor the Mayans realized quite how useful it could be.

In any dynamic, positional number system, a placeholder zero assumes almost unannounced a new guise: it becomes a mathematical "operator" that brings the full power of the system's base to bear. This becomes obvious when we consider the result of adding a placeholder zero to the end of a decimal number string. The number 2,012 becomes 20,120, magically multiplied by the base of 10. We intuitively take advantage of this characteristic whenever we sum two or more numbers, and the total of a column ticks over from 9 to 10. We "carry the one" and leave a zero to ensure the right answer. The simplicity of such algorithms is the source of our system's supple muscularity in manipulating numbers.

We shouldn't blame the Babylonians or Mayans for missing out on such subtlety: various blemishes in their numerical systems made it hard to spot. And so, although they found zero the symbol, they missed zero the number.

Zero is admittedly not an entirely welcome addition to the pantheon of numbers. Accepting it invites all sorts of logical wrinkles that, if not handled with due care and attention, can bring the entire number system crashing down. Adding zero to itself does not result in any increase in its size, as it does for any other number. Multiply any number, however big, by zero and it collapses down to zero. And let's not even delve into what happens when we divide a number by zero.

Classical Greece, the next civilization to handle the concept, was certainly not keen to tackle zero's complexities. Greek thought was wedded to the idea that numbers expressed geometrical shapes; and what shape would correspond to something that wasn't there? It could only be the total absence of something, the void—a concept that the dominant cosmology of the time had banished.

Largely the product of Aristotle and his disciples, this world view saw the planets and stars as embedded in a series of concentric celestial spheres of finite extent. These spheres were filled with an ethereal substance, all centered on Earth and set in motion by an "unmoved mover." It was a picture later eagerly co-opted by Christian philosophy, which saw in the unmoved mover a ready-made identity for God. And since there was no place for a void in this cosmology, it followed that it—and everything associated with it—was a godless concept.

Eastern philosophy, rooted in ideas of eternal cycles of creation and destruction, had no such qualms. And so the next great staging post in zero's journey was not to Babylon's west, but to its east. It is found in *Brahmasphutasiddhanta*, a treatise on the relationship of mathematics to the physical world written in India around AD 628 by the astronomer Brahmagupta.

Brahmagupta was the first person we see treating numbers as purely abstract quantities separate from any physical or geometrical reality. This allowed him to consider unorthodox questions that the Babylonians and Greeks had ignored or dismissed, such as what happens when you subtract from one number a number of greater size. In geometrical terms this is a nonsense: what area is left when a larger area is subtracted? Equally, how could I ever have sold or bartered more goats than I had in the first place? As soon as numbers become abstract entities, however, a whole new world of possibilities is opened up—the world of negative numbers.

The result was a continuous number line stretching as far as you could see in both directions, showing both positive and negative numbers. Sitting in the middle of this line, a distinct point along it at the threshold between the positive and negative worlds, was *sunya*, the nothingness. Indian mathematicians had dared to look into the void—and a new number had emerged.

It was not long before they unified this new number with zero the symbol. While a Christian Syrian bishop writes in 662 that Hindu mathematicians did calculations "by means of nine signs," an inscription of dedication at a temple in the great medieval fort at Gwalior, south of Delhi in India, shows that two centuries later the nine had become ten. A zero—a squashed-egg symbol recognizably close to our own—had been incorporated into the canon, a full member of a dynamic positional number system running from 0 to 9. It marked the birth of the purely abstract number system now used throughout the world, and soon spawned a new way of doing mathematics to go with it: algebra.

News of these innovations took a long time to filter through to Europe. It was only in 1202 that a young Italian, Leonardo of Pisa—better remembered as Fibonacci—published a book, *Liber Abaci*, in which he presented details of the Arabic counting system he had encountered on a journey to the Mediterranean's southern shores, and demonstrated the superiority of this notation over the abacus for the deft performance of complex calculations.

While merchants and bankers were quickly convinced of the Hindu–Arabic system's usefulness, the governing authorities were less enamored. In 1299, the rulers of the Italian city of Florence banned the use of the Hindu–Arabic numerals, including zero. They considered the ability to inflate a number's value hugely simply by adding a digit on the end—a facility not available in the then-dominant,

A brief history of nothing

Zero is crucial for mathematics, but it has taken thousands of years for its importance to be recognized

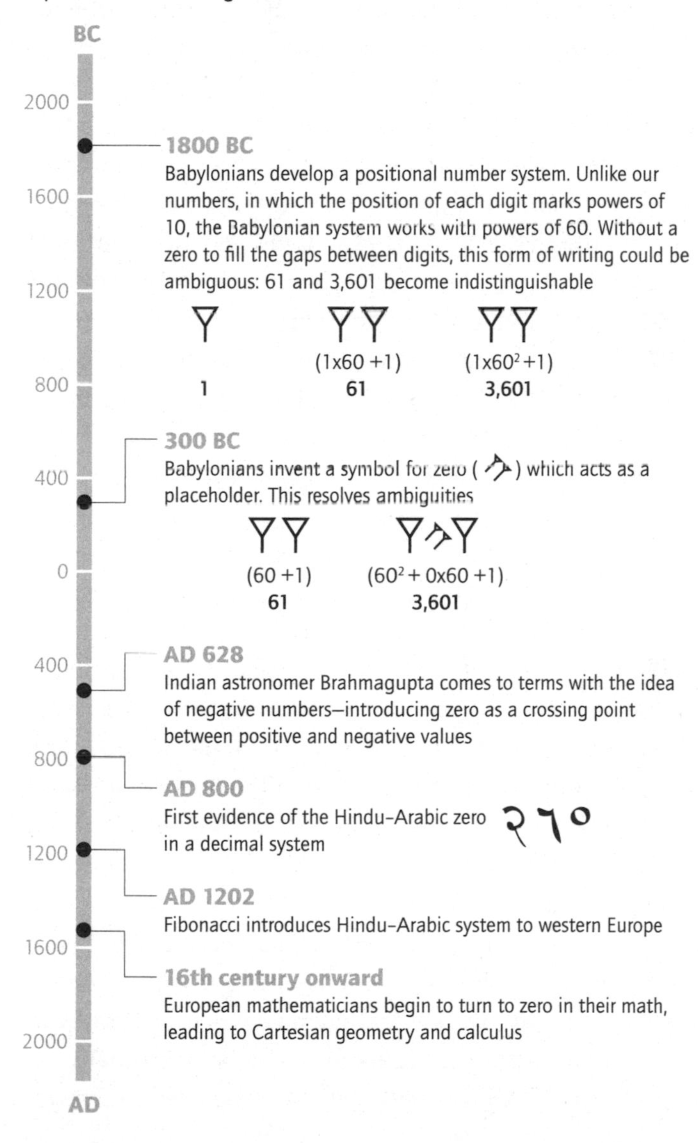

non-positional system of Roman numerals—to be an open invitation to fraud.

Zero the number had an even harder time. Schisms, upheavals, reformation and counter-reformation in the church meant a continuing debate as to the worth of Aristotle's ideas about the cosmos, and with it the orthodoxy or otherwise of the void. Only the Copernican revolution—the crystal-sphere-shattering revelation that the Earth moves around the sun—began, slowly, to shake European mathematics free of the shackles of Aristotelian cosmology from the 16th century onward.

By the 17th century, the scene was set for zero's final triumph. It is hard to point to a single event that marked it. Perhaps it was the advent of the coordinate system invented by the French philosopher and mathematician René Descartes. His Cartesian system married algebra and geometry to give every geometrical shape a new symbolic representation with zero, the unmoving heart of the coordinate system, at its center. Zero was far from irrelevant to geometry, as the Greeks had suggested: it was essential to it. Soon afterward, the new tool of calculus showed that you had first to appreciate how zero merged into the infinitesimally small to explain how anything in the cosmos could change its position at all—a star, a planet, a hare overtaking a tortoise. Zero was itself the prime mover.

Thus a better understanding of zero became the fuse of the scientific revolution that followed. Subsequent events have confirmed just how essential zero is to mathematics and all that builds on it. Looking at zero sitting quietly in a number today, and primed with the concept from a young age, it is equally hard to see how it could ever have caused so much confusion and distress. A case, most definitely, of much ado about nothing.

To read more about zero, go to "Zero, zip, zilch" on page 118.

Heal thyself

We tend to think of medicine as being all about pills and potions recommended to us by another person—a doctor. But science is starting to reveal that for many conditions another ingredient could be critical to the success of these drugs, or perhaps even replace them. That ingredient is nothing more than your own mind. Jo Marchant counts six ways to raid your built-in medicine cabinet.

Better believe it

"I talk to my pills," says Dan Moerman, an anthropologist at the University of Michigan-Dearborn. "I say, 'Hey guys, I know you're going to do a terrific job.'"

That might sound eccentric, but based on what we've learned about the placebo effect, there is good reason to think that talking to your pills really can make them do a terrific job. The way we think and feel about medical treatments can dramatically influence how our bodies respond.

Simply believing that a treatment will work may trigger the desired effect even if the treatment is inert—a sugar pill, say, or a saline injection. For a wide range of conditions, from depression to Parkinson's, osteoarthritis and multiple sclerosis, it is clear that the placebo response is far from imaginary. Trials have shown measurable changes such as the release of natural painkillers, altered neuronal firing patterns, lowered blood pressure or heart rate and boosted immune response, all depending on the beliefs of the patient.

It has always been assumed that the placebo effect only works if people are conned into believing that they are getting an actual active drug. But now it seems this may not be true. Belief in the placebo effect itself—rather than a particular drug—might be enough to encourage our bodies to heal.

In a recent study, Ted Kaptchuk of Harvard Medical School in Boston and his colleagues gave people with irritable bowel syndrome an inert pill. They told them that the pills were "made of an inert substance, like sugar pills, that have been shown in clinical studies to produce significant improvement in IBS symptoms through mind-body self-healing processes," which is perfectly true. Despite knowing the pills were inert, on average the volunteers rated their symptoms as moderately improved after taking them, whereas those given no pills said there was only a slight change.[1]

"Everybody thought it wouldn't happen," says the study's co-author Irving Kirsch, a psychologist at the University of Hull. He thinks that the key was giving patients something to believe in. "We didn't just say 'here's a sugar pill.' We explained to the patients why it should work, in a way that was convincing to them."

As well as having implications for the medical profession, the study raises the possibility that we could all use the placebo effect to convince ourselves that sucking on a sweet or downing a glass of water, for example, will banish a headache, clear up a skin condition or boost the effectiveness of any drugs that we take. "Our study suggests that might indeed help," says Kirsch. While Moerman talks to his pills, Kirsch recommends visualising the desired improvement and telling yourself that something is going to get better.

Think positive

"Everything's going to be fine." Go on, try to convince yourself, because realism can be bad for your health. Optimists recover better from medical procedures such as coronary bypass surgery, have healthier immune systems and live longer, both in general and when suffering from conditions such as cancer, heart disease and kidney failure.[2]

It is well accepted that negative thoughts and anxiety can make us ill. Stress—the belief that we are at risk—triggers physiological pathways such as the "fight-or-flight" response, mediated by the sympathetic nervous system. These have evolved to protect us from danger, but if switched on long-term they increase the risk of conditions such as diabetes and dementia.

What researchers are now realising is that positive beliefs don't just work by quelling stress. They have a positive effect too—feeling safe and secure, or believing things will turn out fine, seems to help the body maintain and repair itself. A recent analysis of various studies concluded that the health benefits of such positive thinking happen independently of the harm caused by negative states such as pessimism or stress, and are roughly comparable in magnitude.[3]

Optimism seems to reduce stress-induced inflammation and levels of stress hormones such as cortisol. It may also reduce susceptibility to disease by dampening sympathetic nervous system activity and stimulating the parasympathetic nervous system. The latter governs what's called the "rest-and-digest" response—the opposite of fight-or-flight.

Just as helpful as taking a rosy view of the future is having a rosy view of yourself. High "self-enhancers"—people who see themselves in a more positive light than others see them—have lower cardiovascular responses to stress and recover faster, as well as lower baseline cortisol levels.[4]

Some people are just born optimists. But whatever your natural disposition, you can train yourself to think more positively, and it seems that the more stressed or pessimistic you are to begin with, the better it will work.

David Creswell from Carnegie Mellon University in Pittsburgh, Pennsylvania, and his colleagues asked students facing exams to write short essays on times when they had displayed qualities that were important to them, such as creativity or independence. The aim was to boost their sense of self-worth. Compared with a control group, students who "self-affirmed" in this way had lower levels of adrenaline and other fight-or-flight hormones in their urine at the time of their exam.[5] The effect was greatest in those who started off most worried about their exam results.

Trust people

Your attitude toward other people can have a big effect on your health. Being lonely increases the risk of everything from heart attacks to dementia, depression and death, whereas people who are satisfied with their social lives sleep better, age more slowly and respond better to vaccines. The effect is so strong that curing loneliness is as good for your health as giving up smoking, according to John Cacioppo of the University of Chicago, Illinois, who has spent his career studying the effects of social isolation.

"It's probably the single most powerful behavioral finding in the world," agrees Charles Raison of Emory University in Atlanta, Georgia, who studies mind–body interactions. "People who have rich social lives and warm, open relationships don't get sick and they live longer." This is partly because people who are lonely often don't look after themselves well, but Cacioppo says there are direct physiological mechanisms too—related to, but not identical to, the effects of stress.

In 2011, Cacioppo reported that in lonely people, genes involved in cortisol signaling and the inflammatory response were up-regulated, and that immune cells important in fighting bacteria were more active, too. He suggests that our bodies may have evolved so that in situations of perceived social isolation, they trigger branches of the immune system involved in wound healing and bacterial infection. An isolated person would be at greater risk of physical trauma, whereas being in a group might favor the immune responses necessary for fighting viruses, which spread easily between people in close contact.

Crucially, these differences relate most strongly to how lonely people think they are, rather than to the actual size of their social network. That also makes sense from an evolutionary point of view, says Cacioppo, because being among hostile strangers can be just as dangerous as being alone. So ending loneliness is not about spending more time with people. Cacioppo thinks it is all about our attitude to others: lonely people become overly sensitive to social threats and come to see others as potentially dangerous. In a review of previous studies, published in 2010, he found that tackling this attitude reduced loneliness more effectively than giving people more opportunities for interaction, or teaching social skills.[6]

If you feel satisfied with your social life, whether you have one or two close friends or quite a few, there is nothing to worry about. "But if you're sitting there feeling threatened by others and as if you're alone in the world, that's probably a reason to take steps," Cacioppo says.

Meditate

Monks have been meditating on mountaintops for millennia, hoping to gain spiritual enlightenment. Their efforts have probably enhanced their physical health, too.

Trials looking at the effects of meditation have mostly been small, but they have suggested a range of benefits. There is some evidence that meditation boosts the immune response in vaccine recipients and people with cancer, protects against a relapse in major depression, soothes skin conditions and even slows the progression of HIV.

Meditation might even slow the aging process. Telomeres, the protective caps on the ends of chromosomes, get shorter every time a cell divides and so play a role in aging. Clifford Saron of the Center for Mind and Brain at the University of California, Davis, and colleagues showed in 2011 that levels of an enzyme that builds up telomeres were higher in people who attended a three-month meditation retreat than in a control group.[7]

As with social interaction, meditation probably works largely by influencing stress response pathways. People who meditate have lower cortisol levels, and one study showed they have changes in their amygdala, a brain area involved in fear and the response to threat.[8]

One of the co-authors of Saron's study, Elissa Epel, a psychiatrist at the University of California, San Francisco, believes that meditation may also boost "pathways of restoration and health enhancement," perhaps by triggering a release of growth and sex hormones.

If you don't have time for a three-month retreat, don't worry. Imaging studies show that meditation can cause structural changes in the brain after as little as 11 hours of training. Epel suggests fitting in short "mini-meditations" throughout the day, taking a few minutes at your desk to focus on your breathing, for example: "Little moments here and there all matter."

Hypnotize yourself

Hypnotherapy has struggled for scientific acceptance ever since Franz Mesmer claimed in the 18th century that he could cure all manner of ills with what he termed "animal magnetism." "The whole field is plagued by people who don't feel research is necessary," says Peter Whorwell of the University of Manchester.

Whorwell has spent much of his professional life building a body of evidence for the use of hypnosis to treat just one condition: irritable bowel syndrome. IBS is considered a "functional" disorder—a rather derogatory term used when a patient suffers symptoms but doctors can't see anything wrong. Whorwell felt that his patients, some of whom had such severe symptoms they were suicidal, were being let down by the medical profession. "I got into hypnosis because the conventional treatment of these conditions is abysmal."

Whorwell gives patients a brief tutorial on how the gut functions, then gets them to use visual or tactile sensations—the feeling of warmth, for example—to imagine their bowel working normally. It seems to work—IBS is the only condition for which hypnosis is recommended by the UK's National Institute for Health and Clinical Excellence. Despite this, Whorwell still has trouble convincing doctors to prescribe it. "We've produced a lot of incontrovertible research," he says. "Yet people are still loath to agree to it."

Part of the problem is that it isn't clear exactly how hypnosis works. What is clear is that when hypnotized, people can influence parts of their body in novel ways. Whorwell has shown that under hypnosis, some IBS patients can reduce the contractions of their bowel, something not normally under conscious control.[9] Their bowel lining also becomes less sensitive to pain.

Hypnosis probably taps into physiological pathways

similar to those involved in the placebo effect, says Kirsch. For one thing, the medical conditions that the two can improve are similar, and both are underpinned by suggestion and expectation—in other words, believing in a particular outcome. The downside is that some people do not respond as strongly to hypnosis as others.

Most clinical trials involving hypnosis are small, largely because of a lack of funding, but they suggest that hypnosis may help pain management, anxiety, depression, sleep disorders, obesity, asthma and skin conditions such as psoriasis and warts.[10] Finding a good hypnotherapist can be tricky, as the profession is not regulated, but hypnotising yourself seems to work just as well. "Self-hypnosis is the most important part," says Whorwell.

Know your purpose

In a study of 50 people with advanced lung cancer, those judged by their doctors to have high "spiritual faith" responded better to chemotherapy and survived longer. More than 40 percent were still alive after three years, compared with less than 10 percent of those judged to have little faith.[11] Are your hackles rising? You're not alone. Of all the research into the healing potential of thoughts and beliefs, studies into the effects of religion are the most controversial.

There are thousands of studies purporting to show a link between some aspect of religion—such as attending church or praying—and better health. Religion has been associated with lower rates of cardiovascular disease, stroke, blood pressure and metabolic disorders, better immune functioning, improved outcomes for infections such as HIV and meningitis, and lower risk of developing cancer.

Critics of these studies, such as Richard Sloan of Columbia University Medical Center in New York, point out that many of them don't adequately tease out other factors. For instance, religious people often have lower-risk lifestyles and churchgoers tend to enjoy strong social support, and seriously ill people are less likely to attend church. Nonetheless, a 2009 analysis of studies in the area concluded, after trying to control for these factors, that "religiosity/ spirituality" does have a protective effect, though only in healthy people.[12] The authors warned that there might be a publication bias, though, with researchers failing to publish negative results.

Even if the link between religion and better health is genuine, there is no need to invoke divine intervention to explain it. Some researchers attribute it to the placebo effect—trusting that some deity or other will heal you may be just as effective as belief in a drug or doctor.[13] Others, like Paolo Lissoni of San Gerardo Hospital in Milan, who did the lung-cancer study mentioned above, believe that positive emotions associated with "spirituality" promote beneficial physiological responses.

Yet others think that what really matters is having a sense of purpose in life, whatever it might be. Having an idea of why you are here and what is important increases our sense of control over events, rendering them less stressful. In Saron's three-month meditation study, the increase in levels of the enzyme that repairs telomeres correlated with an increased sense of control and an increased sense of purpose in life. In fact, Saron argues, this psychological shift may have been more important than the meditation itself.

He points out that the participants were already keen meditators, so the study gave them the chance to spend three months doing something important to them. Spending more time doing what you love, whether it's gardening or voluntary work, might have a similar effect on health. The big news from the study, Saron says, is "the profound impact of having the opportunity to live your life in a way that you find meaningful."

To read more about the power of nothing, go to "Placebo power" on page 55.

2

Mysteries

The desire to solve mysteries or understand anomalies in the world around us is the driving force of science. Puzzles such as why apples fall downward and why liquids get hot when you stir them enough have led us to important insights. One of the delightful or, depending on your viewpoint, frustrating things about the world's mysteries is that dispelling one often reveals others. So here are four mysteries, unsolved or partly solved, that will hopefully lead to new understanding—and almost certainly to new riddles.

The day time began

In the first chapter we discovered how the universe developed in the big bang from an energy-packed blob the size of a pea. But what happened before that point—how did that nascent cosmos emerge from nothing? Physical evidence to answer this question is non-existent and our understanding is confused by a cloud of metaphysical issues and misconceptions. To cut through the fog, here's physicist Paul Davies.

Can science explain how the universe began? Even suggestions to that effect have provoked an angry and passionate response from many quarters. Religious people tend to see the claim as a move to finally abolish God the Creator. Atheists are equally alarmed, because the notion of the universe coming into being from nothing looks suspiciously like the creation, ex nihilo, of Christianity. The general sense of indignation was well expressed by writer Fay Weldon. "Who cares about half a second after the big bang?" she railed in 1991 in a scathing newspaper attack on scientific cosmology. "What about the half a second before?" What indeed. The simple answer is that, in the standard picture of the cosmic origin, there was no such moment as "half a second before."

To see why, we need to examine this standard picture in more detail. The first point to address is why anyone believes the universe began at a finite moment in time. How do we know that it hasn't simply been around for ever? Most cosmologists reject this alternative because of the severe problem of the second law of thermodynamics. Applied to the universe as a whole, this law states that the cosmos is on a one-way slide toward a state of maximum disorder, or entropy. Irreversible changes, such as the gradual consumption of fuel by the sun and stars, ensure that the universe must eventually "run down" and exhaust its supplies of useful energy. It follows that the universe cannot have been drawing on this finite stock of useful energy for all eternity.

Direct evidence for a cosmic origin in a big bang comes from three observations. The first, and most direct, is that the universe is still expanding today. The second is the existence of a pervasive heat radiation that is neatly explained as the fading afterglow of the primeval fire that accompanied the big bang. The third strand of evidence is the relative abundances of the chemical elements, which can be correctly accounted for in terms of nuclear processes in the hot dense phase that followed the big bang.

But what caused the big bang to happen? Where is the center of the explosion? Where is the edge of the universe? Why didn't the big bang turn into a black hole? These are some of the questions that bemused members of the audience always ask whenever I lecture on this topic. Though they seem pertinent, they are in fact based on an entirely false picture of the big bang. To understand the correct picture, it is first necessary to have a clear idea of what the expansion of the universe entails. Contrary to popular belief, it is not the explosive dispersal of galaxies from a common center into the depths of a limitless void. The best way of viewing it is to imagine the space between the galaxies expanding or swelling.

The idea that space can stretch, or be warped, is a central prediction of Einstein's general theory of relativity, and has been well enough tested by observation for all professional cosmologists to accept it. According to general relativity, space-time is an aspect of the gravitational field. This field manifests itself as a warping or curvature of space-time, and when it comes to the large-scale structure of the universe, such warping occurs in the form of space being stretched with time.

A helpful, albeit two-dimensional, analogy for the expanding universe is a balloon with paper spots stuck to the surface. As the balloon is inflated, so the spots, which play the role of galaxies, move apart from each other. Note that it is the surface of the balloon, not the volume within, that represents the three-dimensional universe.

Now, imagine playing the cosmic movie backward, so that the balloon shrinks rather than expands. If the balloon were perfectly spherical (and the rubber sheet infinitely thin), at a certain time in the past the entire balloon would shrivel to a speck. This is the beginning.

Translated into statements about the real universe, I am describing an origin in which space itself comes into existence at the big bang and expands from nothing to form a larger and larger volume. The matter and energy content of the universe likewise originates at or near the beginning, and populates the universe everywhere at all times. Again, I must stress that the speck from which space emerges is not located in anything. It is not an object surrounded by emptiness. It is the origin of space itself, infinitely compressed. Note that the speck does not sit there for an infinite duration. It appears instantaneously from nothing and immediately expands. This is why the question of why it does not collapse to a black hole is irrelevant. Indeed, according to the theory of relativity, there is no possibility of the speck existing through time because time itself begins at this point.

This is perhaps the most crucial and difficult aspect of the big bang theory. The notion that the physical universe came into existence with time and not in time has a long history, dating back to St Augustine in the 5th century. But it took Einstein's theory of relativity to give the idea scientific respectability. The key feature of the theory of relativity is that space and time are part of the physical universe, and not merely an unexplained background arena in which the universe happens. Hence the origin of the physical universe must involve the origin of space and time too.

But where could we look for such an origin? Well, the theory of relativity permits space and time to possess a variety of boundaries or edges, technically known as singularities. One type of singularity exists in the center of a black hole. Another corresponds to a past boundary of space and time at the big bang. The idea is that, as you move backward in time, the universe becomes more and more compressed and the curvature or warping of spacetime escalates without limit, until it becomes infinite at the singularity. Very roughly, it resembles the apex of a cone, where the fabric of the cone tapers to an infinitely sharp point and ceases. It is here that space and time begin.

Once this idea is accepted, it is immediately obvious that the question "What happened before the big bang?" is meaningless. There was no such epoch as "before the big bang," because time began with the big bang. Unfortunately, the question is often answered with the bald statement "There was nothing before the big bang,"and this has caused yet more misunderstandings. Many people interpret "nothing" in this context to mean empty space, but as I have been at pains to point out, space simply did not exist prior to the big bang.

Perhaps "nothing" here means something more subtle, like pre-space, or some abstract state from which space emerges? But again, this is not what is intended by the word. As Stephen Hawking has remarked, the question "What lies north of the North Pole?" can also be answered by "nothing," not because there is some mysterious Land of Nothing there, but because the region referred to simply does not exist. It is not merely physically, but also logically, non-existent. So too with the epoch before the big bang.

In my experience, people get very upset when told this. They think they have been tricked, verbally or logically. They suspect that scientists can't explain the ultimate origin of the universe and are resorting to obscure and dubious concepts like the origin of time merely to befuddle their detractors. The mindset behind such outraged objection is understandable: our brains are hardwired for us to think in terms of cause and effect. Because normal physical causation takes place within time, with effect following cause, there is a natural tendency to envisage a chain of causation stretching back in time, either without any beginning, or else terminating in a metaphysical First Cause, or Uncaused Cause, or Prime Mover. But cosmologists now invite us to contemplate the origin of the universe as having no prior cause in the normal sense, not because it has an abnormal or supernatural prior cause, but because there is simply no prior epoch in which a preceding causative agency—natural or supernatural—can operate.

Nevertheless, cosmologists have not explained the origin of the universe by the simple expedient of abolishing any preceding epoch. After all, why should time and space have suddenly "switched on"? One line of reasoning is that this spontaneous origination of time and space is a natural consequence of quantum mechanics. Quantum mechanics is the branch of physics that applies to atoms and subatomic particles, and it is characterized by Werner Heisenberg's uncertainty principle, according to which sudden and unpredictable fluctuations occur in all observable quantities. Quantum fluctuations are not caused by anything—they are genuinely spontaneous and intrinsic to nature at its deepest level.

For example, take a collection of uranium atoms undergoing radioactive decay due to quantum processes in their nuclei. There will be a definite time period, the half-life, after which half the nuclei should have decayed. But according to Heisenberg it is not possible, even in principle, to predict when a particular nucleus will decay. If you ask, having seen a particular nucleus decay, why the decay event happened at that moment rather than some other, there is no deeper reason, no underlying set of causes, that explains it. It just happens.

The key step for cosmogenesis is to apply this same idea not just to matter, but to space and time as well. Because space-time is an aspect of gravitation, this entails applying quantum theory to the gravitational field of the universe. The application of quantum mechanics to a field is fairly routine for physicists, though it is true that there are special technical problems associated with the gravitational case that have yet to be resolved. The quantum theory of the origin of the universe therefore rests on shaky ground.

In spite of these technical obstacles, one may say quite generally that once space and time are made subject to quantum principles, the possibility immediately arises of space and time "switching on," or popping into existence, without the need for prior causation, entirely in accordance with the laws of quantum physics.

If a big bang is permitted by the laws of physics to

happen once, such an event should be able to happen more than once. In recent years a growing posse of cosmologists has proposed models of the universe involving many big bangs, perhaps even an infinite number of them. In the model known as eternal inflation there is no ultimate origin of the entire system, although individual "pocket universes" within the total assemblage still have a distinct origin. The region we have been calling "the universe" is viewed as but one "bubble" of space within an infinite system of bubbles. In what follows I shall ignore this popular elaboration and confine my discussion to the simple case where only one bubble of space—a single universe—pops into existence.

Even in this simple case, the details of the cosmic birth remain subtle and contentious, and depend to some extent on the interrelationship between space and time. Einstein showed that space and time are closely interwoven, but in the theory of relativity they are still distinct. Quantum physics introduces the new feature that the separate identities of space and time can be "smeared" or "blurred" on an ultra-microscopic scale. In a theory proposed in 1982 by Hawking and American physicist Jim Hartle, this smearing implies that, closer and closer to the origin, time is more and more likely to adopt the properties of a space dimension, and less and less likely to have the properties of time. This transition is not sudden, but blurred by the uncertainty of quantum physics. Thus, in Hartle and Hawking's theory, time does not switch on abruptly, but emerges continuously from space. There is no specific first moment at which time starts, but neither does time extend backward for all eternity.

Unfortunately, the topic of the quantum origin of the universe is fraught with confusion because of the publicity given to a preliminary, and in my view wholly unsatisfactory, theory of the big bang based on an instability of the quantum vacuum. According to this alternative theory, first mooted by Edward Tryon in 1973, space and time are eternal, but matter is not. It suddenly appears in a pre-existing and unexplained void due to quantum vacuum fluctuations. In such a theory, it would indeed involve a serious misnomer to claim that the universe originated from nothing: a quantum vacuum in a background space-time is certainly not nothing!

However, if there is a finite probability of an explosive appearance of matter, it should have occurred an infinite time ago. In effect, Tryon's theory and others like it run into the same problem of the second law of thermodynamics as most models of an infinitely old universe.

Of course, this attempt to explain the origin of the universe is based on an application of the laws of physics. This is normal in science: one takes the underlying laws of the universe as given. But when tangling with ultimate questions, it is only natural that we should also ask about the status of these laws. One must resist the temptation to imagine that the laws of physics, and the quantum state that represents the universe, somehow exist before the universe. They don't—any more than they exist north of the North Pole. In fact, the laws of physics don't exist in space and time at all. They describe the world, they are not "in" it. However, this does not mean that the laws of physics came into existence with the universe. If they did—if the entire package of physical universe plus laws just popped into being from nothing—then we cannot appeal to the laws to explain the origin of the universe. So to have any chance of understanding scientifically how the universe came into existence, we have to assume that the laws have an abstract, eternal character. The alternative is to shroud the origin in mystery and give up.

It might be objected that we haven't finished the job by baldly taking the laws of physics as given. Where did those laws come from? And why those laws rather than some other set? This is a valid objection. I have argued that we must eschew the traditional causal chain and focus instead on an explanatory chain, but inevitably we now confront the logical equivalent of the First Cause—the beginning of the chain of explanation.

In my view it is the job of physics to explain the world based on lawlike principles. Scientists adopt differing attitudes to the metaphysical problem of how to explain the principles themselves. Some simply shrug and say we must just accept the laws as a brute fact. Others suggest that the laws must be what they are from logical necessity. Yet others propose that there exist many worlds, each with differing laws, and that only a small subset of these universes possess the rather special laws needed if life and reflective beings like ourselves are to emerge. Some skeptics rubbish the entire discussion by claiming that the laws of physics have no real existence anyway—they are merely human inventions designed to help us make sense of the physical world. It is hard to see how the origin of the universe could ever be explained with a view like this.

In my experience, almost all physicists who work on fundamental problems accept that the laws of physics have some kind of independent reality. With that view, it is possible to argue that the laws of physics are logically prior to the universe they describe. That is, the laws of physics stand at the base of a rational explanatory chain, in the same way that the axioms of Euclid stand at the base of the logical scheme we call geometry. Of course, one cannot prove that the laws of physics have to be the starting point of an explanatory scheme, but any attempt to explain the world rationally has to have some starting point, and for most scientists the laws of physics seem a very satisfactory one. In the same way, one need not accept Euclid's axioms as the starting point of geometry; a set of theorems like Pythagoras's would do equally well. But the purpose of science (and mathematics) is to explain the world in as simple and economic a fashion as possible, and Euclid's axioms and the laws of physics are attempts to do just that.

In fact, it is possible to quantify the degree of compactness and utility of these explanatory schemes using a branch of mathematics called algorithmic information theory. Obviously, a law of physics is a more compact description of the world than the phenomena that it describes. For example, compare the succinctness of Newton's laws with the complexity of a set of astronomical tables for the positions of the planets. Although as a consequence of Kurt Gödel's famous incompleteness theorem of logic, one cannot prove a given set of laws, or mathematical axioms, to be the most compact set possible, one can investigate mathematically whether other logically self-consistent sets of laws exist. One can also determine whether there is anything unusual or special about the set that characterizes the observed universe as opposed to other possible universes. Perhaps the observed laws are in some sense an optimal set, producing maximal richness and diversity of physical forms. It may even be that the existence of life or mind relates in some way to this specialness. These are open questions, but I believe they form a more fruitful meeting ground for science and theology than dwelling on the discredited notion of what happened before the big bang.

To continue reading about cosmology, go to "Pathways to cosmic oblivion" on page 216.

Placebo power

The placebo effect has been called "the power of nothing." It exerts a healing influence mediated by the mind. But if you think you know how it works, you may be in for a surprise: recent studies have revealed a deeply mysterious effect that threatens the credibility of modern medicine. Michael Brooks finds out why.

It seemed like a good idea until I saw the electrodes. Dr Luana Colloca's white coat offered scant reassurance. "Do you mind receiving a series of electric shocks?" she asked.

I could hardly say no—after all, this was why I was here. Colloca's mentor, Fabrizio Benedetti of the University of Turin, had invited me to come and experience their placebo research first hand. Colloca strapped an electrode to my forearm and sat me in a reclining chair in front of a computer screen. "Try to relax," she said.

First, we established my pain scale by determining the mildest current I could feel, and the maximum amount I could bear. Then Colloca told me that, before I got another shock, a red or a green light would appear on the computer screen.

A green light meant I would receive a mild shock. A red light meant the shock would be more severe, like the jolt you get from an electric fence. All I had to do was rate the pain on a scale of 1 to 10, mild to severe.

After 15 minutes—and what seemed like hundreds of shocks—the experiment ended with a series of mild shocks. Or so I thought, until Colloca told me that these last shocks were in fact all severe.

I had felt the "electric fence" jolts as a series of gentle taps on the arm because my brain had been conditioned to anticipate low pain whenever it saw a green light—an example of the placebo effect.

Benedetti watched the procedure with a smile on his face. He was not sure his team could induce a placebo response in me if I knew I was about to be deceived. As it turns out, I succumbed, hook, line and sinker.

Such is the power of placebo. This was once thought to be a simple affair, involving little more than the power of positive thinking. Make people believe they are receiving good medical care—with anything from a sugar pill to a kindly manner—and in many cases they begin to feel better without any further medical intervention.

However, Benedetti and others are now claiming that the true nature of placebo is far more complex. The placebo effect, it turns out, can lead us on a merry dance. Drug trials, Benedetti says, are particularly problematic.

"An ineffective drug can be better than a placebo in a standard trial," says Benedetti.

The opposite can also be true, as Ted Kaptchuk of Harvard Medical School in Boston points out. "Often, an active drug is not better than placebo in a standard trial, even when we can be confident that the active drug does work," he says.

Some researchers are so taken aback by the results of their studies that they are calling for the very term "placebo" to be scrapped. Others suggest the latest findings undermine the very foundations of evidence-based medicine. "Placebo is ruining the credibility of medicine," Benedetti says.

How did it come to this? After all, the foundation of evidence-based medicine, the clinical trial, is meant to rule out the placebo effect.

If you're testing a drug such as a new painkiller, it's supposed to work like this. First, you recruit the test subjects. Then you assign each person to one of two groups randomly, to ensure there is no bias in the way members of the groups are chosen. One group gets the painkiller, the other gets a dummy treatment. Then, you might think, all you have to do is compare the two groups.

It's not that simple, though, because this is where the placebo problem kicks in. If people getting an experimental painkiller expect it to work, it will work to some extent—just as seeing a green light reduced the pain I felt when shocked. If those in the control group know they're getting a dummy pill whereas those in the other group know they're getting the "real" drug, the experimental painkiller might appear to work better than the dummy when in fact the difference between the groups is entirely due to the placebo effect.

So it's crucial not to tell the subjects what they are getting. Those running the trial should not know either, so they cannot give anything away, creating the gold standard of clinical trials, the double-blind randomized controlled trial. This does not eliminate the placebo effect, but should make it equal in both groups. According to conventional wisdom, in a double-blind trial any "extra" effect in the group given the real drug must be entirely down to the drug's physical effect.

Benedetti, however, has shown this is not necessarily true. His early work in this area involved an existing painkiller called a CCK-antagonist. First, he performed a standard double-blind randomized controlled trial. As you would expect, the CCK-antagonist performed better than the placebo. Standard interpretation: the CCK-antagonist is an effective painkiller.

Now comes the mind-boggling part. When Benedetti gave the same drug to volunteers without telling them what he was doing, it had no effect.[1] "If it were a real painkiller, we should expect no difference compared to the routine overt administration," he says. "What we found is that the covert CCK-antagonist was completely ineffective in relieving pain."

Benedetti's team has since shown that the combination of a patient's expectation plus the administration of the CCK-antagonist stimulates the production of natural painkilling endorphins. It has been known since 1978 that the placebo effect alone can relieve pain in this way. What Benedetti has uncovered, however, is a far more complex interaction between a drug and the placebo effect. His work suggests that the CCK-antagonist is not actually a painkiller in the conventional sense, but more of a "placebo amplifier"—and the same might be true of many other drugs.

"We can never be sure about the real action of a drug," says Benedetti. "The very act of administering a drug activates a complex cascade of biochemical events in the patient's brain." A drug may interact with these expectation-activated molecules, confounding the interpretation of results.

This could be true of some rather famous—and profitable—substances. Benedetti has found that diazepam, for instance, doesn't reduce anxiety in patients after an operation unless they know they are taking it. The placebo effect is required in order for it to be effective. It's not yet clear if this is also true of diazepam's other effects.

Even with drugs that do have direct effects independently of patients' expectations, the strength of these effects can be influenced by expectation. If you don't tell people that they are getting an injection of morphine, you have to inject at least 12 milligrams to get a painkilling effect, whereas if you tell them, far lower doses can make a difference.

Such findings prove that we need to change the way trials are done, Benedetti says. He thinks this is true of all placebo-controlled trials, not just those involving conditions in which placebos can have a strong effect, such as those on pain.

The alternatives include Benedetti's hidden treatment approach, where participants are not always told when they are getting a drug, and the "balanced placebo design," in which you tell some people they got the drug when they actually got the placebo and vice versa.

These approaches are a great way of teasing apart true drug effects from placebo, says Franklin Miller of the US National Institutes of Health. The problem is the degree of deception involved. "There is no way we're going to be able to do clinical trials that involve deceiving patients about what they are getting. I don't see that as a useable method," he says.

Colloca disagrees. With hidden treatments, a patient might not know when the drug infusion starts and ends, but they know that a drug will be given. Therefore, she argues, there is full informed consent.

For Kaptchuk, the issue is not just teasing apart drug effects from placebo; it's the very notion that only treatments that are better than placebo have any value. "It's never enough to just test against placebo," he says.

In a study published in 2008, his team compared three "treatments" for irritable bowel syndrome.[2] One group got sham acupuncture and lots of attention. The second group also got sham acupuncture, but no attention. A third group of patients just got left on a "waiting list."

Patients in both sham acupuncture groups did better than those kept on the fake waiting list. However, the group who had felt listened to and consulted about their symptoms, feelings and treatments reported an improvement that was equivalent to the "positive" trial results for drugs commonly used to treat irritable bowel syndrome—drugs that are supposed to have been proved better than placebo. Does this finding mean the drugs should not have been approved, even though patients are better off with either drugs or placebo than no treatment at all?

This study shows how a placebo can be boosted by combining factors that contribute to its effect. And all sorts of factors can be involved. Even word of mouth can help, Colloca says, such as learning that a treatment has worked for others.

Conditioning through repetition, as in the process I went through, is another important factor. "Many trials use the repeated administration of drugs, thus triggering learning mechanisms that lead to increased placebo responsiveness," Benedetti says.

This is yet another reason to change clinical trials, he argues. It could explain, for instance, why the placebo effect appears to be growing stronger in clinical trials, causing problems for drug companies attempting to prove their products are better than placebo.

The issue isn't just about disentangling the effect of a placebo from that of a drug. It's also about harnessing its power. For instance, Colloca thinks the conditioning effect could be exploited to reduce doses of painkilling drugs with potentially dangerous side effects.[3]

The problem with trying to exploit the placebo effect, says Miller, is that the term means very different things to different people. Many doctors don't believe placebo has any effect other than to placate those demanding some kind of treatment. "People say it's noise, or nothing, or something just to please the patient," Miller says.

Those involved with clinical trials, by contrast, tend to overestimate the power of placebo. Consider the way trials are carried out. If the people in the control group—the ones who receive the placebo—get better, it's almost always attributed to the placebo effect. But in fact, there are many other reasons why those in the control group can improve. Many conditions get better all by themselves given enough time, for instance. To distinguish between the apparent effect of a placebo and its actual effect, you have to compare a placebo treatment with no treatment at all, as in the irritable bowel study.

In an article published in 2008, Miller and Kaptchuk argue that the very notion of placebo has became so laden with baggage that it should be ditched.[4] Instead, they suggest that doctors and researchers should think in terms of "contextual healing"—the aspect of healing that is produced, activated or enhanced by the context of the clinical encounter, rather than by the specific treatment given.

Whatever you call it, trying to harness the placebo effect raises tricky ethical issues: can doctors exploit it without lying to patients? Maybe. If my shocking experience is anything to go by, knowing you are getting a placebo does not necessarily stop it working.

"It's a complicated issue, but one that deserves a lot more attention," Miller says. "Finding ethically appropriate ways to tap the use of placebo in clinical practice is where the field needs to be moving."

Doctors, however, are not hanging around waiting for the results of rigorous studies that show whether or not placebos can be used effectively and ethically for specific conditions. Surveys suggest around half of doctors regularly prescribe a placebo and that a substantial minority do so not just to get patients out of the consulting room but because they believe placebos produce objective benefits.

Are they doing their patients a disservice? In 2001, Asbjørn Hróbjartsson of the Nordic Cochrane Institute in Stockholm did a meta-analysis of 130 clinical trials that compared the placebo group with a no-treatment group, to reveal the "true" placebo effect. The studies involved around 7,500 patients suffering from about 40 different conditions ranging from alcohol dependence to Parkinson's disease.

The meta-analysis concluded that, overall, placebos have no significant effects. Two years later the team published a follow-up study with data from 11,737 patients. "The results are similar again," Hróbjartsson says. Placebos are overrated and largely ineffective, he concludes, and doctors should stop using them. A third review of 202 trials delivers the same message.[5]

However, if you consider only studies whose outcomes are measured by patients' reports, such as how much pain they feel, placebos do appear to have a small but significant effect, he says. In other words, the placebo effect can make you feel better—even if you aren't actually better.

Does this mean it's not a real effect? Was I deluded when I reported feeling severe shocks as mild ones? "What does 'real' mean in this situation?" responds Hróbjartsson. "My concern is not so much whether effects of placebo are real or not, but whether there is evidence for clinically relevant effects."

Giving patients plenty of TLC is where placebo intervention should end, he thinks. "Most of us working in the field think that's just another way of saying 'Be a good doctor.'"

Colloca and Benedetti think there is scope for doing

much better than that. "We already know that placebos don't work everywhere, therefore the small magnitude of the placebo effect in that meta-analysis is not surprising at all," Benedetti says. "It is as if you wanted to test the effects of morphine in gut disorders, pain, heart diseases, marital discord, depression and such like. If you average the effects of morphine across all these conditions, the outcome would be that overall morphine is ineffective."

The other reason not to take the meta-analyses too seriously is the evidence that placebo can have measurable biochemical effects. The release of painkilling endorphins, for instance, has been confirmed by showing that drugs which block endorphins also block the placebo effect on pain, and by brain scans that "light up" endorphins. Placebos have also been shown to trigger the release of dopamine in people with Parkinson's disease. In 2004, Benedetti demonstrated that, after conditioning, individual neurons in the brains of Parkinson's patients respond to a salt solution in the same way as they do to a genuine drug designed to relieve tremors.

When it comes to the placebo effect, it seems, nothing is simple. We still have a lot to learn about this elusive phenomenon.

For more on the power of nothing, go to "When mind attacks body" on page 133.

Wastes of space?

What have the appendix, wisdom teeth and the coccyx got in common? They all do nothing except occasionally cause pain, right? So why have we got them? One suggestion is that they are remnants of organs that served us well in our evolutionary past. It's a neat idea, but evidence to support it is not always easy to find, as Laura Spinney discovers.

Vestigial organs have long been a source of perplexity and irritation for doctors and of fascination for the rest of us. In 1893, a German anatomist named Robert Wiedersheim drew up a list of 86 human "vestiges," organs "formerly of greater physiological significance than at present." Over the years, the list grew, then shrank again. Today, no one can remember the score. It has even been suggested that the term is obsolete, useful only as a reflection of the anatomical knowledge of the day. In fact, these days many biologists are extremely wary of talking about vestigial organs at all.

This may be because the subject has become a battlefield for those who refuse to accept that evolution exists: the creationist and intelligent design lobbies. They argue that none of the items on Wiedersheim's original list are now considered vestigial, so there is no need to invoke evolution to explain how they lost their original functions. While they are right to question the status of some organs that were formerly considered vestiges, denying the concept altogether flies in the face of the biological facts. While most biologists prefer to steer clear of what they see as a political debate, Gerd Müller, a theoretical biologist from the University of Vienna, is fighting to bring the concept back into the scientific arena. "Vestigiality is an important biological phenomenon," he says.

Part of the problem, says Müller, is semantic: people have come to think of vestigial organs as useless, which is not what Wiedersheim said. In an attempt to clear up the confusion, Müller has come up with a more explicit definition: vestigial structures are largely or entirely functionless as far as their original roles are concerned—though they may retain lesser functions or develop minor new ones. Müller points out that it is useful to know if a given structure is vestigial, both for taxonomic purposes—understanding how different species are related to one another—and for medical reasons, as in the case of an organ that has no obvious use in adults but turns out to be crucial in development.

Nobody doubts that some human structures that were once considered vestigial have proved to be far from redundant in the light of growing medical knowledge. For example, Wiedersheim's original list included such eminently useful structures as the three smallest toes and the valves in veins that prevent blood flowing backward. It also contained several organs we now know to be part of the immune system, such as the adenoids and tonsils, lymphatic tissues that produce antibodies, and the thymus gland in the upper chest, which is important for the production and maturation of T-lymphocytes.

Some of Wiedersheim's vestiges have since been identified as hormone-secreting glands—notably the pituitary at the base of the brain, which regulates homeostasis, and the pineal located deep in the brain, which secretes the hormone melatonin. Melatonin is best known for synchronising the activity of our internal organs, including the reproductive organs, with the diurnal and seasonal cycles, but it is also a potent antioxidant that protects the brain and other tissues from damage, so slowing down the aging process.

Then there is the male nipple. The most showily useless of all human structures would seem to be a lock for continued inclusion on Wiedersheim's list. However, evolutionary biologist Andrew Simons of Carleton University in Ottawa says any claim that it is vestigial is bogus. To be vestigial, an organ or something from which it is derived must have had a function in the first place. "There is no reason to believe that male nipples ever served any function," says Simons. Instead, they persist because all human foetuses share the same basic genetic blueprint and males retain a feature that is useful in females because there is no adaptive cost in doing so.

Natural selection shapes living organisms to survive, Müller points out. Once their survival is ensured, the organism may very well retain non-adaptive or non-functional features, provided they are not a burden. This is one reason why we shouldn't expect anatomical structures to be perfectly adapted to their function (or lack of one), says Simons. It also complicates the identification of truly vestigial structures.

Another problem arises when trying to show that a modern structure has lost function with respect to its ancestral form. Take the appendix, a small worm-like pouch that protrudes from the cecum of the large intestine. It was long thought to be a shriveled-up remnant of some larger digestive organ—primarily because it is a lot less prominent than its counterpart in rabbits, with which it had previously been compared. In 1980, G. B. D. Scott of London's Royal Free Hospital put that assumption to the test. He compared the appendix in different primate species and found the human appendix to be among the largest and most structurally distinct from the cecum. "It develops progressively in the higher primates to culminate in the fully developed organ seen in the gorilla and man," he wrote.

Scott concluded that the appendix is far from functionless in apes and humans. Recent evidence from another quarter seems to support his finding. A study by R. Randal Bollinger and colleagues at Duke University School of Medicine at Durham, North Carolina, found that the human appendix acts as a "safe house" for helpful, commensal bacteria, providing them with a place to grow and, if necessary, enabling them to re-inoculate the gut should it lose its normal microbial inhabitants—for example, as a result of illness.[1] Although people who have had their appendix removed seem to suffer no ill effects, team member Bill Parker points out that the operation is mainly performed in the developed world. "If you lived in a traditional culture, any time before 1800, or in a developing country where they don't have sewer systems, you are going to need your appendix," he says. Parker suspects that far from being vestigial, the specialized appendix evolved out of a cecum that had the more general twin functions of housing good bacteria and aiding digestion.

That would explain Scott's finding. However, there is an alternative explanation that allows for the possibility that the appendix is a vestigial organ. In 1998, evolutionary theorists Randolph Nesse of the University of Michigan, Ann Arbor, and the late George Williams of the State University of New York at Stony Brook argued that while you might expect natural selection to eliminate the annoying human appendix if it could, we might paradoxically be stuck with it.[2] They pointed out that a smaller, thinner appendix would be more likely to become blocked by inflammation and inaccessible to a cleansing blood supply, increasing the risk of life-threatening infections. Their conclusion was that "larger appendixes are thus actually selected for," even though they may no longer have a role.

The jury is still out on the human appendix, but examples from other animals leave no doubt that vestigiality is a real phenomenon. Look no further than the wings on flightless birds for an unequivocal example of a vestige, says paleontologist Gareth Dyke of University College Dublin. The loss of flight in large, ground-living birds happened relatively recently—within the past 50 million years—and usually as a result of the birds being restricted to an island, or because of the loss of terrestrial predators. The ostrich is an extreme case, because its wings have already lost some of the bones that were present in its airborne ancestors. "The feathers too are modified," Dyke says. "They are not flight feathers any more. There's no structure to them. They are just really fluffy, downy feathers."

So, with the benefit of modern scientific knowledge, what are the most convincing examples of vestigial structures in humans? The *New Scientist* top-five list runs as follows: the vomeronasal organ, goose bumps, Darwin's point, the tail bone and wisdom teeth (see "Top five human vestiges" below). There are undoubtedly more: it depends how wide you cast your net, says Müller. Some blood vessels have become reduced in size and function over time, and thinking smaller still, there must be chemical messengers and genes that qualify.

As the late evolutionary biologist Stephen Jay Gould pointed out, nobody ever said evolution was perfect. The existence of something as spectacularly *de trop* as the ostrich wing is only a problem for those who believe in an intelligent designer. On the other hand, the list of vestigial organs should still be considered a work in progress. Anything that appears to be entirely without function is suspicious, says Müller, and probably just waiting to be assigned one. Whether we are talking about useless vestiges or anatomical structures that have taken on a new lease of life, however, it is hard to ignore the evidence that human beings are walking records of their evolutionary past.

Top five human vestiges

Vomeronasal organ

Many animals secrete chemical signals called pheromones that carry information about their gender or reproductive state, and influence the behavior of others. In rodents and other mammals, pheromones are detected by a specialized sensory system, the vomeronasal organ (VNO), which consists of a pair of structures that nestle in the nasal lining or the roof of the mouth. Although most adult humans have something resembling a VNO in their nose, neuroscientist Michael Meredith of Florida State University in Tallahassee has no hesitation in dismissing it as a remnant.

"If you look at the anatomy of the structure, you don't see any cells that look like the sensory cells in other mammalian VNOs," he says. "You don't see any nerve fibers connecting the organ to the brain." He also points to genetic evidence that the human VNO is non-functional. Virtually all the genes that encode its cell-surface receptors—the molecules that bind incoming chemical signals, triggering an electrical response in the cell—are pseudogenes, and inactive.

So what about the puzzling evidence that humans respond to some pheromones? The late Larry Katz and a team at Duke University, North Carolina, found that as well as the VNO, the main olfactory system in mice also responds to pheromones. If that is the case in humans too then it is possible that we may still secrete pheromones to influence the behavior of others without using a VNO to detect them.

Goose bumps

Though goose bumps are a reflex rather than a permanent anatomical structure, they are widely considered to be vestigial in humans. The pilomotor reflex, to give them one of their technical names, occurs when the tiny muscle at the base of a hair follicle contracts, pulling the hair upright. In birds or mammals with feathers, fur or spines, this creates a layer of insulating warm air in a cold snap, or a reason for a predator to think twice before attacking. But human hair is so puny that it is incapable of either of these functions.

Goose bumps in humans may, however, have taken on a minor new role. Like flushing, another thermoregulatory mechanism, they have become linked with emotional responses—notably fear, rage or the pleasure, say, of listening to beautiful music. This could serve as a signal to others. It may also heighten emotional reactions: there is some evidence, for instance, that a music-induced frisson causes changes of activity in the brain that are associated with pleasure.

Darwin's point

Around the sixth week of gestation, six swellings of tissue called the hillocks of Hiss arise around the area that will form the ear canal. These eventually coalesce to form the outer ear. Darwin's point, or tubercle, is a minor malformation of the junction of the fourth and fifth hillocks of Hiss. It is found in a substantial minority of people and takes the form of a cartilaginous node or bump on the rim of their outer ear, which is thought to be the vestige of a joint that allowed the top part of the ancestral ear to swivel or flop down over the opening to the ear.

Technically considered a congenital defect, Darwin's point does no harm and is surgically removed for cosmetic reasons only. However, the genetics behind it tell an interesting tale, says plastic surgeon Anthony Sclafani of the New York Eye and Ear Infirmary. The trait is passed on according to an autosomal dominant pattern, meaning that a child need only inherit one copy of the gene responsible to have Darwin's point. That suggests that at one time it was useful. However, it also has variable penetration, meaning that you won't necessarily have the trait even if you inherit the gene. "The variable penetration reflects the fact that it is no longer advantageous," Sclafani says.

The tail bone

A structure that is the object of reduced evolutionary pressure can, within limits, take on different forms. As a result, one of the telltale signs of a vestige is variability. A good example is the human coccyx, a vestige of the mammalian tail, which has taken on a modified function, notably as an anchor point for the muscles that hold the anus in place. The human coccyx is normally composed of four rudimentary vertebrae fused into a single bone. "But it's amazing how much variability there is at this spot," says Patrick Foye, director of the Coccyx Pain Center at New Jersey Medical School in Newark. Whereas babies born with six fingers or toes are rare, he says, the coccyx can and often does consist of anything from three to five bony segments. What's more, there are more than 100 medical reports of babies born with tails. This atavism arises if the signal that normally stops the process of vertebrate elongation during embryonic development fails to activate on time.

Wisdom teeth

Most primates have wisdom teeth (the third molars), but a few species, including marmosets and tamarins, have none. "These are probably evolutionary dwarfs," says anthropologist Peter Lucas of George Washington University, Washington DC. He suggests that when the body size of mammals reduces rapidly, their jaws become too small to house all their teeth, and overcrowding eventually results in selection for fewer or smaller teeth.[3] This seems to be happening in *Homo sapiens*.

Robert Corruccini of Southern Illinois University in Carbondale says the problem of overcrowding has been exacerbated in humans in the past four centuries as our diet has become softer and more processed. With less wear on molars, jaw space is at an even higher premium, "so the third molars, the last teeth to erupt, run out of space to erupt," he says. Not only are impacted wisdom teeth becoming more common, perhaps as many as 35 percent of people have no wisdom teeth at all, suggesting that we may be on an evolutionary trajectory to losing them altogether.

If you want to read about other things that do nothing, go to "Busy doing nothing" on page 93.

Banishing consciousness

Anesthetics are incredibly valuable drugs. After you've received one, you feel absolutely nothing, which is why they are so useful in surgery. They also show great promise in helping to answer one of humanity's biggest questions—what is consciousness? Linda Geddes finds out more, some of it from personal experience.

I walk into the operating theater feeling vulnerable in a draughty gown and surgical stockings. Two anesthetists in green scrubs tell me to stash my belongings under the gurney and lie down. "Can we get you something to drink from the bar?" they joke, as one deftly slides a needle into my left hand.

I smile weakly and ask for a gin and tonic. None appears, of course, but I begin to feel light-headed, as if I really had just knocked back a stiff drink. I glance at the clock, which reads 10.10 am, and notice my hand is feeling cold. Then, nothing.

I have had two operations under general anesthetic in 12 months. On both occasions I awoke with no memory of what had passed between the feeling of mild wooziness and waking up in a different room. Both times I was told that the anesthetic would make me feel drowsy, I would go to sleep, and when I woke up it would all be over.

What they didn't tell me was how the drugs would send me into the realms of oblivion. They couldn't. The truth is, no one knows.

The development of general anesthesia has transformed surgery from a horrific ordeal into a gentle slumber. It is one of the commonest medical procedures in the world, yet we still don't know how the drugs work. Perhaps this isn't surprising: we still don't understand consciousness, so how can we comprehend its disappearance?

That is starting to change, however, with the development of new techniques for imaging the brain or recording its electrical activity during anesthesia. "In the past five years there has been an explosion of studies, both in terms of consciousness, but also of how anesthetics might interrupt consciousness and what they teach us about it," says George Mashour, an anesthetist at the University of Michigan in Ann Arbor. "We're at the dawn of a golden era."

Consciousness has long been one of the great mysteries of life, the universe and everything. It is something experienced by every one of us, yet we cannot even agree on how to define it. How does the small sac of jelly that is our brain take raw data about the world and transform it into the wondrous sensation of being alive? Even our increasingly sophisticated technology for peering inside the brain has, disappointingly, failed to reveal a structure that could be the seat of consciousness.

Altered consciousness doesn't only happen under a general anesthetic of course—it occurs whenever we drop off to sleep, or if we are unlucky enough to be whacked on the head. But anesthetics do allow neuroscientists to manipulate our consciousness safely, reversibly and with exquisite precision.

It was a Japanese surgeon who performed the first known surgery under anesthetic, in 1804, using a mixture of potent herbs. In the west, the first operation under general anesthetic took place at Massachusetts General Hospital in 1846. A flask of sulphuric ether was held close to the patient's face until he fell unconscious.

Since then, a slew of chemicals have been co-opted to serve as anesthetics, some inhaled, like ether, and some injected. The people who gained expertise in administering these agents developed into their own medical specialty. Although long overshadowed by the surgeons who patch you up, the humble "gas man" does just as important a job, holding you in the twilight between life and death.

Consciousness may often be thought of as an all-or-nothing quality—either you're awake or you're not—but as I experienced, there are different levels of anesthesia (see figure below). "The process of going into and out of general anesthesia isn't like flipping a light switch," says Mashour. "It's more akin to a dimmer switch."

A typical subject first experiences a state similar to drunkenness, which they may or may not be able to recall later, before falling unconscious, which is usually defined as failing to move in response to commands. As

You are feeling sleepy
Losing consciousness under anesthesia is like turning down a dimmer switch

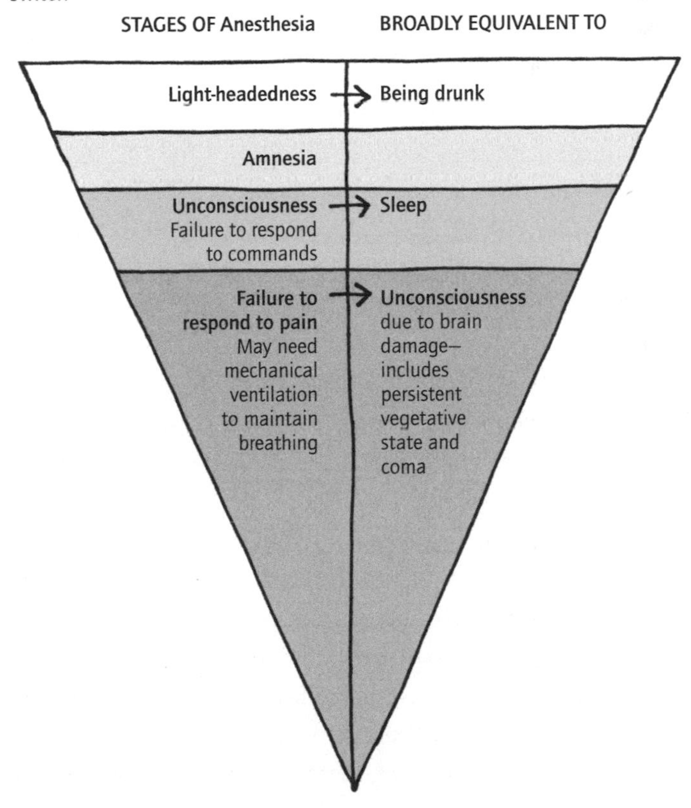

they progress deeper into the twilight zone, they fail to respond to even the penetration of a scalpel—which is the point of the exercise, after all—and at the deepest levels may need artificial help with breathing.

These days anesthesia is usually started off with injection of a drug called propofol, which gives a rapid and smooth transition to unconsciousness, as happened with me. (This is also what Michael Jackson was allegedly using as a sleeping aid, with such unfortunate consequences.) Unless the operation is only meant to take a few minutes, an inhaled anesthetic, such as isoflurane, is then usually added to give better minute-by-minute control of the depth of anesthesia.

So what do we know about how anesthetics work? Since they were first discovered, one of the big mysteries has been how the members of such a diverse group of chemicals can all result in the loss of consciousness. Many drugs work by binding to receptor molecules in the body, usually proteins, in a way that relies on the drug and receptor fitting snugly together like a key in a lock. Yet the long list of anesthetic agents ranges from large complex molecules such as barbiturates or steroids to the inert gas xenon, which exists as mere atoms. How could they all fit the same lock?

For a long time there was great interest in the fact that the potency of anesthetics correlates strikingly with how well they dissolve in olive oil. The popular "lipid theory" said that instead of binding to specific protein receptors, the anesthetic physically disrupted the fatty membranes of nerve cells, causing them to malfunction.

In the 1980s, though, experiments in test tubes showed

that anesthetics could bind to proteins in the absence of cell membranes. Since then, protein receptors have been found for many anesthetics. Propofol, for instance, binds to receptors on nerve cells that normally respond to a chemical messenger called GABA. Presumably the solubility of anesthetics in oil affects how easily they reach the receptors bound in the fatty membrane.

But that solves only a small part of the mystery. We still don't know how this binding affects nerve cells, and which neural networks they feed into. "If you look at the brain under both xenon and propofol anesthesia, there are striking similarities," says Nick Franks of Imperial College London, who overturned the lipid theory in the 1980s. "They must be triggering some common neuronal change, and that's the big mystery."

Many anesthetics are thought to work by making it harder for neurons to fire, but this can have different effects on brain function, depending on which neurons are being blocked. So brain-imaging techniques such as fMRI, which tracks changes in blood flow to different areas of the brain, are being used to see which regions of the brain are affected by anesthetics. Such studies have been successful in revealing several areas that are deactivated by most anesthetics. Unfortunately, so many regions have been implicated that it is hard to know which, if any, are the root cause of loss of consciousness.

But is it even realistic to expect to find a discrete site or sites acting as the mind's "light switch"? Not according to a leading theory of consciousness that has gained ground in the past decade, which states that consciousness is a more widely distributed phenomenon. In this "global workspace" theory, incoming sensory information is first processed locally in separate brain regions without us being aware of it.[1] We only become conscious of the experience if these signals are broadcast to a network of neurons spread through the brain, which then start firing in synchrony.

The idea has recently gained support from recordings of the brain's electrical activity using electroencephalograph (EEG) sensors on the scalp, as people are given anesthesia. This has shown that as consciousness fades there is a loss of synchrony between different areas of the cortex—the outermost layer of the brain important in attention, awareness, thought and memory.[2]

This process has also been visualized using fMRI scans. Steven Laureys, who leads the Coma Science Group at the University of Liège in Belgium, looked at what happens during propofol anesthesia when patients descend from wakefulness, through mild sedation, to the point at which they fail to respond to commands. He found that while small "islands" of the cortex lit up in response to external stimuli when people were unconscious, there was no spread of activity to other areas, as there was during wakefulness or mild sedation.[3]

A team led by Andreas Engel at the University Medical Center in Hamburg has been investigating this process in still more detail by watching the transition to unconsciousness in slow motion. Normally it takes about 10 seconds to fall asleep after a propofol injection. Engel has slowed it down to many minutes by starting with just a small dose, then increasing it in seven stages. At each stage he gives a mild electric shock to the volunteer's wrist and takes EEG readings.

We know that upon entering the brain, sensory stimuli first activate a region called the primary sensory cortex, which runs like a headband from ear to ear. Then further networks are activated, including frontal regions involved in controlling behavior, and temporal regions toward the base of the brain that are important for memory storage.

Engel found that at the deepest levels of anesthesia, the primary sensory cortex was the only region to respond to the electric shock. "Long-distance communication seems to be blocked, so the brain cannot build the global workspace," says Engel, who presented the work at the Society for Neuroscience annual meeting in 2010. "It's like the message is reaching the mailbox, but no one is picking it up."

What could be causing the blockage? Engel has EEG data suggesting that propofol interferes with communication between the primary sensory cortex and other brain regions by causing abnormally strong synchrony between them. "It's not just shutting things down. The communication has changed," he says. "If too many neurons fire in a strongly synchronized rhythm, there is no room for exchange of specific messages."

The communication between the different regions of the cortex is not just one-way; there is both forward and backward signaling between the different areas. EEG studies on anesthetized animals suggest it is the backward signal between these areas that is lost when they are knocked out.

In October 2011, Mashour's group published EEG work showing this to be important in people too. Both propofol and the inhaled anesthetic sevoflurane inhibited the transmission of feedback signals from the frontal cortex in anesthetized surgical patients. The backward signals recovered at the same time as consciousness returned.[4] "The hypothesis is whether the preferential inhibition of feedback connectivity is what initially makes us unconscious," he says.

Similar findings are coming in from studies of people in a coma or persistent vegetative state (PVS), who may open their eyes in a sleep–wake cycle, although remaining unresponsive. Laureys, for example, has seen a similar breakdown in communication between different cortical areas in people in a coma. "Anesthesia is a pharmacologically induced coma," he says. "That same breakdown in global neuronal workspace is occurring."

Many believe that studying anesthesia will shed light on disorders of consciousness such as coma. "Anesthesia studies are probably the best tools we have for understanding consciousness in health and disease," says Adrian Owen of the University of Western Ontario in London, Canada.

Owen and others have previously shown that people in a PVS respond to speech with electrical activity in their brain. More recently he did the same experiment in people progressively anesthetized with propofol. Even when heavily sedated, their brains responded to speech. But closer inspection revealed that those parts of the brain that decode the meaning of speech had indeed switched off, prompting a rethink of what was happening in people with PVS.[5] "For years we had been looking at vegetative and coma patients whose brains were responding to speech and getting terribly seduced by these images, thinking that they were conscious," says Owen. "This told us that they are not conscious."

As for my own journey back from the void, the first I remember is a different clock telling me that it is 10.45 am. Thirty-five minutes have elapsed since my last memory—time that I can't remember, and probably never will.

"Welcome back," says a nurse sitting by my bed. I drift in and out of awareness for a further undefined period, then another nurse wheels me back to the ward, and offers me a cup of tea. As the shroud of darkness begins to lift, I contemplate what has just happened. While I have been asleep, a team of people have rolled me over, cut me open, and rummaged about inside my body—and I don't remember any of it. For a brief period of time "I" had simply ceased to be.

My experience leaves me with a renewed sense of awe for what anesthetists do as a matter of routine. Without really understanding how, they guide hundreds of millions of people a year as close to the brink of nothingness as it is possible to go without dying. Then they bring them safely back home again.

If you want to find out more about feeling nothing, go to "Ride the celestial subway" on page 142.

3

Making sense of it all

Our lives have been transformed by knowledge gleaned from scientific discoveries and the technologies they have prompted. It would be nice to think that these advances emerged from a straightforward path between questions being asked and solutions presenting themselves. But that is rarely the case. Scientists may not even know which questions to ask of a new phenomenon, and technologies are sometimes used for a long time before we realize how they work. Yet successes there are, often with their own built-in surprises. Here are a few.

Out of thin air

The philosopher Bertrand Russell wrote that "almost every serious intellectual advance has had to begin with an attack on some Aristotelian doctrine." A good example of this is Aristotle's insistence that the vacuum could not exist. Only in the 17th century, when people had seriously started to ask questions about the world around them, did a simple experiment overturn this thinking by revealing unequivocally that the void is a real, reproducible thing. Physicist Per Eklund traces the story to the present day.

Among the torrent of scientific ideas and technological advances of the 17th century, four are often singled out as "great inventions" for their profound impact on the way humanity views reality. The telescope helped us to understand the solar system and that Earth is not at its center. The microscope opened up a rich yet previously invisible "microworld" and paved the way for modern medicine, materials technology and much more. The pendulum clock was our first accurate timekeeper and a fundament of modern society.

Perhaps the least well known of the four is the vacuum pump. Its impact is less obvious than the others, possibly because changes in pressure are more difficult to appreciate than images seen through an eyepiece and are not as obviously useful as shared time. What's more, uses for the vacuum were not immediately forthcoming. Only in the past 150 years has the vacuum become critical in shaping our view of matter and the way we live.

Aristotle's idea that "nature abhors a vacuum" is one we might smile at today in a world where vacuums exist in everything from food packaging and vacuum cleaners to the Large Hadron Collider. Yet in 1600, almost 2,000 years after Aristotle lived, his view held firm that a vacuum, or empty space, could not exist. Nature, he had argued, was comprised of four elements (earth, wind, fire and water), and space could not be defined where there was nothing. That view began to change following groundbreaking experiments.

The crucial experiments that challenged the status quo were carried out by the Italians Gasparo Berti in 1640 and Evangelista Torricelli in 1644. In Torricelli's experiment, he filled a meter-long test tube with mercury and turned it upside down in a basin of mercury. The height of mercury in the tube immediately fell to 760 millimeters leaving a gap at the top that could only be a vacuum.

Berti conducted his experiment first, using water instead of mercury, but he was less successful in convincing contemporary natural philosophers of his case. Torricelli is therefore usually given the historic honor of "discovering" the vacuum. In doing so he also unwittingly invented the mercury barometer. A few years later, the Frenchman Blaise Pascal made the first pressure measurement using Torricelli's set-up—the mercury column rose or fell depending on the atmospheric pressure. And at about the same time, Otto van Guericke in Germany constructed an "air pump," a manually controlled wooden piston pump, rather like a bicycle pump in reverse, that we know today as the first vacuum pump.

One might think that now vacuums were philosophically feasible and technologically possible, there would be a race to reach ever-lower pressures. But this did not happen. Progress halted for some 200 years. Apart from a few inquisitive natural philosophers, nobody could think what to do with the vacuum, so it became something of a sideshow. By 1850, the base pressure that could be reached had been reduced to the vicinity of 1 or 2 millibar, compared with the 6 mbar attained by Robert Boyle in 1660. (To get some feeling for these pressures, see box, "What is low pressure like?")

What is low pressure like?

It is much more difficult to comprehend a low pressure than, say, a small distance. Vacuums encountered in our daily lives are still close to atmospheric pressure. For example, 1 bar is the atmospheric pressure at sea level; a household vacuum cleaner achieves around 750 millibar; and the pressure outside a plane at 10,000 meters altitude is about 250 mbar. But even very simple vacuum pumps reach pressures of the order of 1 mbar.

Picture a milk carton with a volume of 1 liter—that's one-thousandth of a cubic meter—filled with air at atmospheric pressure. A pressure of 1 mbar corresponds to the same amount of air in 1 cubic meter. If the cube is 10 meters on a side, the pressure is 10^{-3} mbar, and 10^{-6} mbar would correspond to a 100-meter-sided cube, larger than any skyscraper.

Two of the most important units of measurement for pressure are named after the earliest contributors to vacuum technology. Evangelista Torricelli's column of mercury gave rise to the logical pressure unit: 1 millimeter of mercury (mm Hg), or a "Torr." The SI unit for pressure is the Pascal (Pa), after Blaise Pascal who first measured pressure using Torricelli's set-up. In vacuum science and technology both mbar and Torr are more extensively used than the SI unit. 1 mbar = 0.75 Torr = 100 Pa. Normal air pressure is 1,013 mbar or 760 Torr.

The first major pressure drop came in the mid-1800s, with a mercury pump still based on Torricelli's principle. This pump, designed by the German Heinrich Geissler in 1855 and then improved on by his compatriot Herman Sprengel, was a hand-driven mercury displacement pump. It forced mercury to drop from a capillary tube at the top of a flask into another at the bottom. Consecutive drops trapped between them tiny amounts of air from the flask, driving down its pressure. Sprengel's apparently simple set-up improved the attainable pressure by a breathtaking six orders of magnitude to reach 10^{-5} mbar.

This improvement is an early example of research driven by need. The pumps grew out of the requirements of industry and scientists such as Thomas Edison in the second half of the 19th century. Edison needed a way to stop the filaments in his incandescent light bulbs from burning up. Creating a vacuum inside the bulb solved the problem, and the Sprengel pump found rapid use in his factory.

Likewise, studies of gas discharges in glass tubes, and what we today call plasmas, would not have been possible without Sprengel's pump. These led to exploration of "cathode rays," which culminated in the discovery of the electron in 1898. Wilhelm Röntgen used vacuum tubes to create the first X-rays, as did Heinrich Hertz in his discovery of the photoelectric effect. These experiments also dispelled the foggy picture of the nature of a vacuum, enhancing the notion that all matter—including rarefied gases at low pressure—is made up of atoms and molecules.

Largely thanks to these advances, the foundations of modern vacuum physics were laid down in the early 1900s. The first of two key breakthroughs was Danish physicist Martin Knudsen's categorization of the mechanisms of gas flow at reduced pressure through long narrow tubes. Vacuums today are defined according to his three regimes: a "molecular" flow regime, where the gas is so diluted that molecules interact with each other individually rather than collectively; an intermediate regime; and a "viscous" regime at pressures within a few orders of magnitude of atmospheric, where the gas behaves like a regular fluid. Each regime is governed by fundamentally different physics, and this insight was critical in developing modern vacuum science.

The second breakthrough was Wolfgang Gaede's pump, created in 1905. It employed a mechanism similar to Sprengel's in trapping air between drops of mercury, but used a rotating mechanism, which allowed it to be mechanically driven. Gaede's pump was an order of magnitude faster than Sprengel's and was the first to require a "backing pump," which kept the outlet below atmospheric pressure. He had invented a pump whose ultimate achievable pressure was so low that the pump could not be connected directly to atmospheric pressure. Backing pumps are now common.

Gaede continued to work on novel pump designs, and in 1915 he presented the first diffusion pump, a type that is still a reliable industrial workhorse today. Despite its title, the pump works not by diffusion but by transferring momentum. Gaede directed a high-speed jet of mercury vapor to propel gas molecules toward the pump's exhaust, like a game of atomic billiards. Modern-day diffusion pumps use various oils instead of mercury, but the design is essentially the same. The oil diffusion pump came a decade after Gaede's invention and could reach 10^{-7} or 10^{-8} mbar of base pressure.

In the second half of the 20th century, two main needs drove the development of vacuum technology. Thriving high-tech industry, especially making semiconductors for electronics, needed reliable, fast, clean vacuum pumps and chambers for large-scale processing. Yet the main driver for the lowest attainable pressure was the growth of "big science." The competition for nuclear weapons and space exploration that characterized the Cold War, and the particle accelerators used in peaceful research into nuclear and particle physics, both demanded high vacuums.

A key invention from this period is the 1957 turbomolecular pump, or "turbo" as it is affectionately known to vacuum scientists. Similar in appearance to a jet engine, it "bats" gas molecules toward the pump's exhaust with rapidly rotating rotor blades. The velocity of the blades is critical: to generate the required momentum, a typical turbo runs at 30,000 to 75,000 revolutions a minute. A turbo can reach pressures as low as about 10^{-10} mbar, but it cannot go lower because the lighter gases, especially hydrogen, never gain the necessary momentum to be expelled.

Will we ever be able to reach a perfect vacuum? The answer is simple: no. Even if a "macroscopic" volume contained no particles, it would still not be truly "empty" because it would remain subject to quantum fluctuations, dark energy, and other quantum-mechanical phenomena. I think the question should rather be rephrased as "will we ever be able to reach a vacuum that, for all practical intents and purposes, would constitute an ideal vacuum?" This can be done, and most likely has already been done.

In a technological vacuum on Earth, the practical limits are much higher than in outer space. No wall or seal is perfect, and permeation of molecules, especially hydrogen, will always result in there being some particles present. This is especially problematic because most ultra-high-vacuum systems use turbos, which are inefficient at pumping hydrogen. For the most advanced applications in surface science such as scanning tunneling microscopy and large-scale facilities such as synchrotrons and particle accelerators, this will remain insufficient.

To improve, the base pressure can be pushed down by cryopumping and gettering. Instead of pumping the gas out of the vacuum system, gettering and cryopumping capture the gas molecules inside the system—sticking them to the vacuum chamber itself. A getter is a highly reactive material, such as titanium, that attracts and adsorbs, or binds, any gas left in a vacuum. Coating exposed areas inside a vacuum chamber with a getter can reduce the pressure generated by a turbo pump by as much as an extra two orders of magnitude.

Only cryopumping, however, has the potential to reach an "ultimately low" pressure. As the name suggests, cryopumping works by cooling exposed areas so they adsorb or "freeze out" any remaining gas molecules. Most cryopumps are cooled by liquid nitrogen and are used in industrial applications to reach similar pressures to those achieved by diffusion pumps. But cryopumping can in principle be used to reach much lower pressures. Just how much lower was suggested by Canadian physicist Peter Hobson's thought experiment on cryopumping about 40 years ago. He assumed an ideal situation in which the vacuum container would be a sealed half-liter glass flask, which would then be immersed in liquid helium at a temperature just 4 degrees above absolute zero. By extrapolating the adsorption curves of gases on glass to the liquid helium temperature, he concluded that "it is quite practical to create a pressure of 10^{-33} [mbar]." That is, a pressure of some 20 orders of magnitude lower than in any existing vacuum system at the time.

There is a big conceptual problem with this set-up, however. One gas particle in a half-liter corresponds to a pressure of about 10^{-20} mbar, so how could we possibly end up at 10^{-33} mbar? This is a neat illustration of the dangers of extrapolation, since achieving that pressure in half a liter would require splitting one molecule into 10^{13} parts!

Nevertheless, the idea of freezing out gases works and has been used at large-scale facilities such as CERN to store antimatter. When antimatter comes into contact with ordinary matter, they annihilate each other with an "explosion," so antiparticles must be kept away from ordinary particles. This can be achieved by guiding antiparticles, such as antiprotons, into a vacuum container that is then enclosed and immersed in liquid helium.

Antimatter can be stored for months at a time under such conditions. Although there is no way to measure the actual pressure in the container, one can calculate that the particle density must be lower than about 100 atoms per cubic centimeter, which corresponds to 10^{-16} mbar, to avoid contact between particles and antiparticles. To my mind, this is the ultimate test of a vacuum. The pressure in this vacuum is so low that practically speaking it is a close approximation of a "perfect" vacuum.

Reaching such pressures needs costly and advanced equipment. Commercially, reasonable results can be achieved with cheaper and less complex set-ups. In most practical situations, obtaining a better vacuum is not a goal in itself. It only needs to be as good as is required for the purpose—productivity and cost are more important. Throughout the history of the vacuum, such commercial and technological considerations have been dominant, and there's little doubt they will remain so into the future.

To carry on reading about the vacuum, go to "Into the void" on page 110.

Busy doing nothing

By any measure some animals are lazy: hanging or lying around all day doing diddly-squat. Sloths are well known for it, and snakes come a close second. But why do they do nothing? When scientists investigated this question, it became clear that these critters have no option. Jonathan Knight finds out why.

After a hard day's work, the journey home, cooking, washing up and putting the kids to bed, it's time to collapse into an armchair for five minutes. And in those few snatched moments, you might well envy the unbelievably easy life of some vertebrates. Take the sloth: it sits motionless for hours up a tree in the rainforest canopy. Or the giant python, which lies around for months waiting for its next meal and then rests in the bushes for weeks doing nothing but digesting. Bliss.

Well, not quite. Research into the behavior and metabolism of such seemingly shiftless animals shows that doing nothing has nothing to do with taking it easy. These animals are operating on the very edge of survival where doing nothing is essential for staying alive. Even more surprisingly, the metabolism of some of the most immobile creatures may be working as hard as a racehorse on the big day.

Mark Chappell, a biologist at the University of California, Riverside, has a particular interest in the energy consumption of animals that live in extreme environments. While working in the Antarctic a couple of decades ago, he discovered that Adélie penguin chicks are not as idle as they seem. Other than occasional brief bouts of begging food from their parents, these juveniles stay fixed to the same spot on the ice for weeks. But when Chappell measured the metabolism of these little birds, he got a shock.

He placed chicks in small sealed chambers with a monitored air supply to find out how fast they were using up oxygen. Oxygen consumption correlates directly with metabolic rate: the higher the demands made on cells, the more oxygen they need to burn glucose and produce energy. Chicks with empty stomachs have a metabolic rate of 1 milliliter of oxygen per gram of body weight per hour.

What really surprised Chappell was that the metabolic rate of freshly fed chicks was twice that figure. Such an increase in metabolic rate is unusually high among warm-blooded animals.

When resting, humans have a metabolic rate of about 0.3 milliliters of oxygen per gram of body weight per hour. Light exercise, such as walking, can double this, whereas sprinting can raise it as much as tenfold over short periods. But, at best, digestion drives our metabolism to only about 50 percent above its resting rate. The rates for most mammals are similar. So a penguin chick digesting its dinner is actually working its metabolism as hard as a human on a brisk walk.

Exercise consumes energy mostly through the working of muscles, but the costs of digestion are more diverse. With the penguin chick's high-protein diet, about half the energy goes on moving the food along the gut, producing digestive enzymes to break down the food, and pumping the resulting molecules into cells in the gut wall. The other half is used within these cells to reassemble amino acids from the food back into proteins.

But why do penguin chicks expend so much energy on these activities? At this early stage of life, penguins have one objective: to grow as quickly as possible. Chicks make good snacks for skuas, predatory gulls that harass the penguins. In this environment, a small, weak chick is a dead chick. So building up body mass fast is vital for survival, and rapid digestion helps this along. If the chicks bolt their food, they can go back to their parents more frequently for more.

So a souped-up metabolic link to the digestive system helps the chicks build up their bodies quickly. This is why idle ways are so useful: energy wasted on movement cannot be converted into body mass. And, as any couch potato knows, doing nothing is a great way to grow fast.

Chappell found still more surprises about the balance between metabolism, exercise and digestion when he studied another bird that lives much closer to home. House wrens nest in tree holes throughout North America. Their chicks, which hatch blind and helpless, sit still for two weeks, converting whatever the parents feed them into bone, muscle, fat and feathers. In that time, they increase their mass nearly tenfold.

To look at their metabolic capabilities, Chappell put baby house wrens in a sealed chamber with a supply of oxygen and a meter to show how much oxygen they used. He tried a number of different conditions, studying well-fed chicks, hungry chicks and hungry chicks that were prodded to keep them moving around in the nest. His results show that during the first eight days nothing the birds did was more strenuous than digesting food.

A six-day old nestling, for example, had a resting rate of 1 milliliter of oxygen per gram of body weight per hour. In its most energetic burst of activity, it could raise that rate by only half again. But by simply sitting still and digesting, a chick could double this rate and then some—an increase even larger than that of the penguin chicks.

Like humans and other mammals, the chicks' parents can double their metabolic rate only when they physically exert themselves. By contrast, the chicks are adapted to draw on this extra energy only when digesting food. They are lazy little growth machines designed to work as hard at digesting food as their parents are at bringing it. Yet after about eight days of doing nothing, Chappell found that the house wren chicks' metabolism flipped so they could consume energy faster during physical activity than during digestion.

In an even more remarkable example of working hard at doing nothing, the Burmese python stays completely still for weeks at a time. Yet the metabolic abilities of this critter rival those of a racehorse going flat out. In his labs at the University of California, Los Angeles, and later at the University of Alabama in Tuscaloosa, Stephen Secor has spent years measuring the rate of oxygen consumption of young Burmese pythons while they are digesting a meal or fasting. The more they eat, he finds, the faster their metabolism.

At a push a small python can eat five rats at a sitting—equal to the snake's own body mass. Snakes that eat this much can increase their metabolism a stunning 44-fold in less than a day. "It looks like they are just sitting there, but they are really huffing and puffing," he says. As the meal is digested, which for big meals can take a couple of weeks, the snakes gradually throttle back again. The only other animal ever measured with a metabolic rate running so far above its resting rate is a thoroughbred horse at full gallop.

Of course, in absolute terms the snake has a much lower metabolic rate than the horse—about an eleventh of its rate, in fact. The snake starts with the very low resting rate of about 0.032 milliliters of oxygen per gram of body weight per hour. Like all cold-blooded animals, a python does not have to maintain a constant body temperature the way mammals and birds do, so it expends much less energy at rest. And a python also feeds so infrequently that to conserve energy it shuts down its gut in between meals.

Digestive tracts are very expensive to maintain, mainly because the cells in contact with the food and digestive juices constantly die and slough off. Replacing them takes energy. But the python stops the digestive juices flowing and temporarily interrupts the cells' replacement cycle. Its gut actually deflates along its length. "This is like turning off a car in a traffic jam to save gas," says Secor.

Secor found that other python organs tighten their

belts in lean times, too. The liver, kidney and heart all shrink gradually as the belly empties. But within a few days of feeding they can grow by up to 50 percent. The gall bladder is the only organ found to lose weight after a meal as it empties stored-up bile into the gut.

These metabolic adaptations are specifically suited to predators that wait to ambush large, rare prey. For the python, which may go for months without finding a meal, keeping still is a matter of life and death. Moving about would not yield enough extra food to justify the expense, because prey is scarce, and would run away if chased, or both. So a python that constantly went hunting would probably die of starvation.

The Burmese python is not alone in adapting to a niche in which the costs of motion outweigh the benefits, says Brian McNab, emeritus professor of biology at the University of Florida and author of *The Physiological Ecology of Vertebrates: A View from Energetics,* which explores the role of energy consumption in evolution. The Texas blind salamander lives in lightless caves where the only food is detritus that washes through in streams or leaks through the roof. The only animal that could survive here is one with a very low energy consumption.

So the salamander is by necessity a sedentary creature. But it doesn't just *do* nothing to survive—it also sees nothing. McNab argues that the creature is descended from lizards that could see, but that the evolutionary pressure to save energy was so strong that it deprived the salamander of its sight.

It is possible that the animal's ability to see evolved away through lack of pressure to maintain it: chance mutations in genes controlling vision could have stopped them working and wouldn't have harmed the salamander's prospects because it didn't need to be able to see. But McNab argues there was probably a selective pressure to drive it away. It takes a great deal of energy to maintain vision, he says, because there is a rapid turnover of cells in the retina and cornea. So keeping the eyes when they were of no use would have been terribly wasteful. "What they are really doing," he says, "is saving energy."

Compared with a dark cave, you might think that the lush rainforests of the Amazon contain an inexhaustible supply of tasty comestibles. And yet among the branches live some remarkably low-energy animals. One of them is the three-toed sloth, a creature that sits motionless for so long that algae grow in its fur.

Even at its peak of activity, climbing a tree in what looks like slow motion, the sloth's metabolic rate never rises above 0.48 milliliters of oxygen per gram of body weight per hour—three times its resting rate. It simply can't spare any more energy. Although the sloth can eat all the leaves it wants, the energy inside the leaf is very hard to get at. It is mostly bound up in cellulose, which takes more energy to break down than the simpler carbohydrates in fruits and the proteins in insects.

What's worse, as a defense against the vast array of insects in the canopy, rainforest leaves have evolved defenses to make them difficult to digest. They contain high concentrations of tannins, which bind to the leaves' proteins and prevent them from being broken down in the animal's gut. This means that much of the energy in the leaf is unavailable. Most leaves also contain powerful toxins such as alkaloids that take energy to inactivate.

In the rainforest, low- and high-energy animals live side by side. Howler monkeys can often be seen hanging listlessly from trees, while capuchins swing playfully from branch to branch stuffing themselves with fruit. Kenneth Glander of Duke University in Durham, North Carolina, has studied New World primates for more than four decades. He says the smaller, lighter capuchins easily outmaneuver the howlers, which weigh nearly twice as much at between 4 and 7 kilograms.

In a sense, the capuchins are creating an extreme environment for the howlers by grabbing all the best food. So when capuchins are around, howlers make do with leaves and slow their metabolism down just as the sloths do. Sometimes, they hang still for hours. Because they must spend so much more energy on processing the leaves in their gut, howlers must do nothing just to survive. "In most cases," says Glander, "howlers are operating on a minimal margin of error." If they were more active, they would probably die of starvation.

So the next time you sit down to put your feet up, remember this: while the chance to do nothing may seem like an evolutionary bonus that humans missed out on, all the research into vertebrate loafers shows that being built to do nothing is about survival—and comes with a penalty clause. What's good for the Texas blind salamander could be bad news for you.

To read more about things that do nothing, go to "Putting the idle to work" on page 183.

The hole story

It's a curious thought that just about anything that contains a silicon chip owes its existence to the movement of, well, nothing. Richard Webb tells us why.

The sound of New Year's Eve celebrations drifting up from the Palace Theater did not distract William Shockley. Nor did the few scattered revellers straying through Chicago's snow-covered streets below. Rarely a mingler, Shockley had more important things on his mind. Barricaded in his room in the art-deco opulence of the Bismarck Hotel, he was thinking, and writing.

Eight days earlier, on December 23, 1947, John Bardeen and Walter Brattain, two of Shockley's colleagues at Bell Laboratories in Murray Hill, New Jersey, had unveiled a device that would change the world: the first transistor. Today, shrunk to just nanometers across and carved into beds of silicon, these electrical on-off switches mass in their billions on every single computer chip. Without them, there would be no processing of the words, sounds and images that guide our electronic lives. There would be no smartphone, router, printer, home computer, server or internet. There would be no information age.

Bardeen and Brattain's device, a rather agricultural construction of semiconductor, gold-enwrapped polystyrene and a spaghetti twist of connecting wires, did not look revolutionary, and it would have taken a seer to foretell the full changes it would bring. Even so, those present that December at Bell Labs knew they had uncovered something big. In Shockley's words, the transistor was a "magnificent Christmas present." Magnificent, but for one thing: no one knew quite how it worked.

Holed up in his Chicago hotel, Shockley needed to change that. As head of Bell Labs' solid-state physics group, he had been the intellectual driving force behind the transistor, yet Bardeen and Brattain had made the crucial breakthrough largely without him. To reclaim the idea as his own, he needed to go one better.

That meant getting to grips with a curious entity that seemed to control the transistor's inner workings. Its existence had been recognized two decades earlier, but its true nature had eluded everyone. For good reason: it was not there.

Transistors—both Bardeen and Brattain's original and those that hum away in computer processors today—depend on the qualities of that odd half-breed of material known as a semiconductor. Sitting on the cusp of electrical conduction and insulation, semiconductors sometimes let currents pass and sometimes resolutely block their passage.

By the early 20th century, some aspects of this dual personality were well documented. For example, the semiconductor galena, or lead sulphide, was known under certain circumstances to form a junction with a metal through which current traveled in only one direction. That had made it briefly popular in early wireless receivers, where a filigree metal probe—a "cat's whisker"—was tickled across a crystal of galena to find the contact that would transform a rapidly oscillating radio signal into a pulsing direct current that could power a speaker.

This process had to be repeated afresh each time a

radio receiver was switched on, which made tuning a time-consuming and sometimes infuriating business. This was symptomatic of all semiconductors' failings. There seemed little rhyme or reason in their properties; a slight change in temperature or their material make-up could tip them from conduction to insulation and back again. It was tempting to think their caprices might be tamed to make reliable, reproducible electrical switches, but no one could see how.

And so in the radio receivers and telephone and telegraph systems of the 1920s and 1930s—such as those operated by Bell Labs' parent company, AT&T—vacuum tubes came to reign supreme. They worked by heating an electrode in a vacuum and applying electric fields of varying strength to the stream of electrons emitted, thus controlling the size of the current reaching a second electrode at the far side. Bulky, failure-prone and power-hungry though they were, vacuum tubes were used as switches and amplifying "repeaters" to hoist fading signals out of a sea of static on their long transcontinental journeys.

Even as they did, however, the seeds of their demise and semiconductors' eventual triumph were being sown. In 1928 Rudolph Peierls, a young Berlin-born Jew, was working as a student of the great pioneer of quantum physics, Werner Heisenberg, in Leipzig. The convolutions of history would later make Peierls one of the UK's most respected physicists, and pit him against his mentor in the race to develop the first atomic bomb. At the time, though, he was absorbed by a more niggling problem: why were electrical currents in some metals deflected the wrong way when they hit a magnetic field?

To Peierls, the answer was obvious. "The point [was] you couldn't understand solids without using the quantum theory," he recalled in a 1977 interview.[1] Just as quantum theory dictates that electrons orbiting an atom can't have just any old energy, but are confined to a series of separate energy states, Peierls showed that within a solid crystal, electrons are shoe-horned into "bands" of allowed energy states. If one of these bands had only a few occupied states, electrons had great freedom to move, and the result was a familiar electron current. But if a band had only a few vacant states, electron movement would be restricted to the occasional hop into a neighboring empty slot. With most electrons at a standstill, these vacancies would themselves seem to be on the move: mobile "absences of electron" acting for all the world like positive charges—and moving the wrong way in a magnetic field.

Peierls never gave these odd non-entities a name. It was Heisenberg who gave them their slightly off-hand moniker: *Löcher*—or "holes." And there things rested. The holes were, after all, just a convenient fiction. Electrons were still doing the actual conducting—weren't they?

Although Peierls's band calculations were the germ of a consistent, quantum-mechanical way of looking at how electrical conduction happened, no one quite joined up the dots at the time. It was ten years before the rumblings of war would begin to change that.

Radar technology, which involves bouncing radio waves off objects to determine their distance and speed, would become crucial to Allied successes in the latter stages of the Second World War. But radar presented a problem. If the equipment were to fly on bombing missions, it needed to be compact and lightweight. Vacuum tubes no longer cut the mustard. Might the long-neglected semiconductors, for all their failings, be a way forward?

In 1940, a team at Bell Labs led by engineer Russell Ohl was exploring that possibility by attempting to tame the properties of the semiconductor silicon. At the time, silicon's grouchy and intermittent conduction was thought to be the result of impurities in its crystal structure, so Ohl and his team set about purifying it. One day, a glitch in the purification process produced a silicon rod with a truly bizarre conducting character. One half acted as if dominated by negatively charged carriers: electrons. The other half, though, seemed to contain moving positive charges.

That was odd, but not half as odd as what happened when you lit up the rod. Left to its own devices, the imbalanced silicon did nothing at all. Shine a bright light on it, however, and it flipped into a conducting state, with current flowing from the negative to the positive region.

A little more probing revealed what was going on. Usually, a silicon atom's four outer electrons are all tied up in bonds to other atoms in the crystal. But on one side of Ohl's rod, a tiny impurity of phosphorus with its five outer electrons was creating an excess of unattached electrons. On the other, a small amount of boron with just three electrons was causing an electron deficit (see figure overleaf).

Peierls's holes had suddenly found a role. When kicked into action by the light, electrons were spilling over from the region of their excess to fill the holes in the electron structure introduced by the boron. However passively, it was the presence of an absence of electrons that was

Holes on the march
In silicon (Si) crystals, all electrons are bound. But add boron (B) atoms, which have one fewer binding electrons, and "holes" are created. Electrons leap between these holes to generate a current (at top). The "holes" effectively flow in the opposite direction (at bottom)

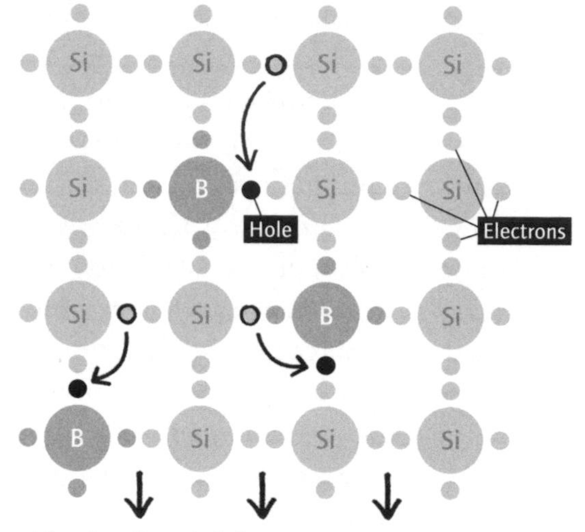

Direction of travel of electrons, or negative charge

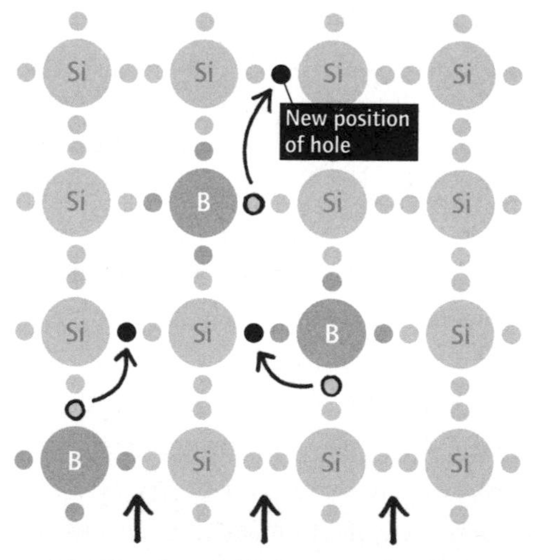

Direction of travel of holes, or positive charge

causing the silicon rod's unique behavior. Ohl named his discovery the positive-negative or "p-n" junction, owing to its two distinct areas of positive and negative charge carriers. Its property of converting light energy into electric current made it, incidentally, the world's first photovoltaic cell.

It was a few years before Shockley got wind of Ohl's breakthrough. Already a senior member of Bell Labs' physics team before the war, he had been taken in a very different direction by the hostilities, becoming head of the US navy's anti-submarine warfare operations research unit. After resurfacing in 1945 to lead Bell's solid-state physics division, it did not take Shockley long to spot the p-n junction's potential.

He was fascinated by the thought that, by adding a metal contact above a junction, you might use an external electric field instead of light to control the current across it. In a sufficiently thin layer of n- or p-type silicon, he reasoned, the right sort of voltage would make electrons or holes swarm toward the contact, providing extra carriers of charge that would boost the current flow across the junction. Varying the voltage would vary the current. The result would be an easily controllable, low-power, small-scale amplifier that would smash the vacuum tube out of sight. That was truly a prospect to pique the interest of Shockley's paymasters.

His first attempts to realize the dream, though, were unsuccessful. "Nothing measurable, no measurable results," he noted of an early failure. "Quite mysterious." And with his mind now on the broad sweep of Bell Labs' solid-state research, Shockley was obliged to leave further investigations to two highly qualified subordinates: Bardeen, a thoughtful theorist, and Brattain, an inveterate tinkerer.

It proved a frustrating chase, and it was a classic combination of experimental acumen and luck that led the pair to success—plus Bardeen's spur-of-the-moment decision to abandon silicon for its slightly more predictable semiconducting sister germanium. This finally produced the right sort of amplification effect, boosting the power of input signals, sometimes by a factor of hundreds. The magnificent Christmas present was unwrapped.

Just one thing didn't add up: the current was moving through the device in the wrong direction. Although the germanium slab had n-type material at the top, it appeared to be positive charges making the running. The puzzlement is almost palpable in Brattain's lab-book entry for December 8, 1947: "Bardeen suggests that the surface field is so strong that one is actually getting p-type conduction near the surface," he wrote. It was a mental block that stopped Bardeen and Brattain understanding the fruits of their labors.

No doubt they would have, given time. But in his Chicago hotel room that New Year's Eve, Shockley stole a march on his colleagues. There was a way out of the impasse, he realized, and he did the first hurried calculations to firm up his case.

If a hole were merely the absence of an electron, then electrons and holes could hardly co-exist: whenever an electron met a hole, its presence would by definition negate the absence of itself that was the hole. By that measure, the existence of positive charges in a negative region, as Bardeen and Brattain had seemingly observed, was nonsense.

But what if a hole were real, Shockley asked: not just an absence of something, but a true nothing-that-is? What if it were a particle all on its own, with an independent existence just as real as the electron's? If this were true, holes would not need to fear encountering an electron. They could happily co-exist with electrons in areas dominated by them—and that would explain what was going on in the transistor.

It was a daring intellectual leap. In the weeks that followed, Shockley used the idea to develop a transistor that exploited the independence of electrons and holes. This was the "p-n-p" transistor, in which a region of electron excess was sandwiched between two hole-dominated areas. Apply the right voltage, and the resistance of the middle section could be broken down, allowing holes to pass through hostile electron-populated territory without being swallowed up. It also worked in reverse: electrons could be made to flow through a central region given over to holes. This was the principle that came to underpin the workings of commercial transistors in the decades that followed.

The rest, as they say, is history. For Shockley, it was not a happy one. He did not at first tell Bardeen and Brattain of his new course, and even attempted to claim sole patent rights over the first transistor.[2] The relationship between the three men never recovered. By the time they shared the Nobel prize in physics for their discovery in 1956, Shockley had left Bell Labs to form the Shockley Semiconductor Laboratory to capitalize on his transistor alone.

But his high-handed and increasingly paranoid behavior soon led to a mass mutiny from the bright young talents he had hired, such as Gordon Moore and Robert Noyce, who went on to found Intel, which remains the world's largest manufacturer of microchips.

The hole, meanwhile, went from strength to strength. Today you will find it at the heart of not just every computer chip, but every energy-saving LED lightbulb, every laser that reads our CDs and DVDs, and every touchscreen. Modern life has become unimaginable without this curiosity whose nature took two decades to reveal: the nothing that became a something and changed the world.

Into the void

As they delve deeper into a topic, scientists regularly find that things are more complicated than they first thought. Genetic screening, for example, has revealed that what we call breast cancer is in fact many distinct diseases. The same thing has happened in space research. What we call the "vacuum of space" turns out to be not one thing but many. Nigel Henbest charts the voids beyond our atmosphere.

In space, no one can hear you scream. There's nothing out there for sound to travel through. Puncture the skin of the International Space Station, and you'd better seal it off pretty quickly or you'll soon be breathing a lot of nothing, in a pressure so low your blood will boil.

And yet, you only have to look at the Hubble Space Telescope's spectacular pictures of gas clouds in space to realize that emptiness is relative. Outer space is not a perfect vacuum, nor are all cosmic vacuums equal. In some places space is teeming with atoms, relatively speaking, while neighboring regions are much emptier. In recent years, astronomers armed with radio telescopes on the ground and a battery of other kinds of telescope in orbit have been mapping the different vacuums in space. The question is, can they find anywhere in the universe where there really is nothing at all?

It's not far from Earth's cocooned surface to the beginning of the cosmic vacuum. As British astrophysicist Fred Hoyle once remarked, "Space is only a two-hour drive away, if your car could go vertically upward." And during those first two hours, you would pass through more gas than in the trillions of years it would take you to travel the remaining distance to the edge of the universe.

For human beings, the region where the International Space Station orbits at 400 kilometers above sea level is as near a true vacuum as makes no difference. You wouldn't last long there without an air supply and pressure suit. Yet the vacuum is not particularly high on a cosmic scale. The air here is still dense enough for astronauts to see a cosmic St Elmo's fire as their spacecraft rips through the tenuous gas at hypersonic speed.

The residual atmosphere up here creates a slight but potentially lethal wind drag on an orbiting spacecraft. One famous casualty was the early US space station Skylab. Launched in 1973, Skylab was slowed by a braking force from this tenuous gas strong enough to make it spiral downward and burn up only six years later. The

International Space Station is continually boosted to save it from a similar fate.

Earth orbit is sometimes touted as a natural laboratory for studying how things behave when both gravity and atmosphere are absent. In fact, there are residues of both. On the International Space Station, thrusters and moving astronauts produce accelerations equivalent to at least several millionths of Earth's gravity. And the "vacuum" outside is nowhere near as good as can be achieved by the best laboratory pumps back on Earth.

Even so, there's something symbolic about the "vacuum of space." In February 2011, space-walking astronaut Al Drew filled a metal cylinder with "space" outside the International Space Station, and brought it back to Earth. This cosmic Message in a Bottle was designed by the Japanese Space Agency to inspire children, as "a conduit between humans and space, and between this world and the one beyond us."

OK, let's be less romantic for a moment and put these different "vacuums" into figures. The range of gas densities across the universe, from Earth's atmosphere to the most tenuous cosmic gas clouds, encompasses so many orders of magnitude that we'd soon get lost in the trillions and billionths. Instead, think in terms of how far apart the atoms or molecules are: the higher the vacuum, the larger the average separation of the gas particles.

At sea level on Earth, air molecules are jostling so closely that they are typically just a millionth of a millimeter apart—only a few times the size of the molecules themselves. In Earth orbit, nature stretches the separation of the molecules to around an average one hundredth of a millimeter. What's more, orbiting vehicles are moving so fast that what the surrounding atoms lack in number they make up in relative speed, as they collide with the craft at 28,000 kilometers per hour.

But there is a way of putting this speed to use to create a high vacuum. In an experiment called Wake Shield, scientists used a simple satellite to smash all records for a human-made vacuum. Wake Shield was essentially a disc of stainless steel, almost four meters across, shaped rather like a saucepan lid. The shield flew on its own several kilometers away from the space shuttle, its convex side forward. As Wake Shield tore through the surrounding gas atoms, it pushed them aside so rapidly that they didn't have time to diffuse around the back of the satellite. The result was a "wake" of gas atoms behind the steel shield, and a high vacuum at its center.

At this altitude, Wake Shield increased the average distance between gas atoms to one-tenth of a millimeter. In three space shuttle flights in the 1990s, Wake Shield's automated lab grew several thin films of semiconductor in the highest artificial vacuum ever achieved—opening the way to ultra-pure chips of new semiconductors and films of high-temperature superconductors.

Go beyond where the International Space Station flies, and Earth's atmosphere eventually peters out. Compared with planetary atmospheres, the gas in outer space is indeed tenuous. But even so, it is far from empty. No sooner do you rise above Earth's shroud of air than you enter the atmosphere of the sun. The hot gases in the sun's outer layer—the corona—constantly boil away into space in a solar wind that sweeps out past the planets.

The blustery solar wind is racked by gusts from magnetic eruptions on the sun's surface. Although they can light up our skies with magnificent displays of auroras, and disrupt electricity supplies down on Earth, we are talking here about storms in a vacuum. The average density of the solar wind is less even than Wake Shield's vacuum, with atoms here about a centimeter apart from each other.

Somewhere well beyond the orbit of Neptune, the solar wind has thinned out so far that it is matched by the tenuous gas that fills the space between the stars in the Milky Way. Interstellar gas is invisible even to the best optical telescopes, but it has an accomplice that gives it away. Scattered throughout the gas are tiny dust particles that absorb light from anything lying behind. Where the gas and dust are most concentrated, you see dark clouds in silhouette. To the naked eye, the Coal Sack near the Southern Cross is one familiar example of interstellar matter in bulk. More spectacularly, the Hubble Space Telescope has revealed the vast dusky "Pillars of Creation," sculpted by dark dust silhouetted against the luminous gases of the Eagle Nebula.

Even though these pillars look dense—almost solid—they are extremely tenuous. The dust specks that make them dark are only the size of a particle of cigarette smoke. They are spread so diffusely that you would find only one, on average, in a volume the size of St Peter's Basilica in Rome. It is only because space is so huge that the dust particles amass to become an all-obscuring fog.

About half the gas in our galaxy lies in relatively dense clouds like these. The rest is spread more widely.

Although it is invisible to ordinary telescopes, its hydrogen atoms emit telltale radio energy at a wavelength of 21 centimeters. This interstellar gas is a tangle of dense strands and tenuous patches. If you traveled through the interstellar medium from a tenuous region to a neighboring denser strand, the density contrast would be greater than diving from Earth's atmosphere into the sea.

Yet even the denser regions contain atoms a centimeter or more apart. In the tenuous patches, the atoms are another ten times further apart. The difference between the two becomes clear when nature unleashes its ultimate stellar cataclysm—the death of a star in a supernova explosion. Supernovae send out a shock wave that speeds through space in a fireball and shows up brilliantly to telescopes tuned to radio waves and X-rays. The shock wave sweeps up the gas it meets like a snowplow, to form a dense shell.

I have a particular affection for this phenomenon, as it provided my introduction to the various vacuums of space. As a radio astronomer in Cambridge, I was checking out the fine details of the fireball left over from the supernova explosion that was seen from Earth in 1572. It was obvious that the fireball isn't spherical: the expanding shell has been pushed out of shape as it encountered irregularities in the surrounding medium. Where the surrounding gas is thinner, the shock travels further and faster; where the shock hits a denser strand it slows down. Like a vast natural Wake Shield, the expanding shock wave clears out a "vacuum within a vacuum," leaving the gas inside the supernova remnant thousands of times less dense than even the tenuous interstellar gases outside it.

And the passage of the fireball has raised the temperature inside the shell to millions of degrees.

The discovery of hot, tenuous gas within a supernova's fireball was no surprise. But astronomers were puzzled to find signs of similar hot and extremely tenuous gas in parts of our galaxy, the Milky Way, where there was no sign of recent supernova explosions.

Gas this thin is very hard to study: with so little material, evidence becomes more and more difficult to find. The first sign that it exists came from the imprint of its spectral lines on the ultraviolet light from distant stars. Its existence was confirmed by the discovery of faint X-rays given off by the gas. Astronomers have now concluded that half the galaxy's volume is filled with million-degree gas whose individual atoms—in fact ions, as they are stripped of their electrons—are fully 10 centimeters apart.

This hot gas, it seems, has come from thousands of supernova explosions over the aeons, each blowing its own bubble in the surrounding interstellar gas. And as the bubbles grew, they coalesced to produce a galaxy-sized "Swiss cheese"—a network of holes surrounded by denser matter. The hot gas has also burst out of its colder surroundings, and now envelops the Milky Way in a faintly glowing halo.

It's a good vacuum—the best in the galaxy—but it's still not perfect. How about the reaches of space beyond? In general, galaxies are not spread uniformly through space. Most of them are gathered in giant clusters, and here you might expect to find the denser parts of the intergalactic medium. Indeed, X-ray telescopes have picked out pools of gas at temperatures of 100 million degrees or more, bound by the gravity of the galaxies in a cluster. The density of these pools is similar to that of the gas in the most tenuous parts of our galaxy, with a distance between ions of about 10 centimeters.

Astronomers are only beginning to probe the intergalactic regions where matter is spread even more thinly. It's too much to hope that the gas here will emit any significant amount of radiation. But that does not mean it is undetectable. The atoms may reveal themselves by absorbing radiation coming from bright beacons beyond. And nature has provided astronomers with suitable light sources: distant and extremely bright galaxies called quasars.

Intergalactic gas is pulled and pummeled by the gravity of galaxies and the enigmatic "dark matter," which is invisible to conventional telescopes yet is thought to make up about 80 percent of all matter. As a result, it should be heated to around a million degrees: a state that astronomers call "warm-hot intergalactic medium," or WHIM for short. As radiation from distant quasars passes through, the WHIM should imprint a dark pattern of absorption lines in the X-ray part of the spectrum.

Using results from two orbiting X-ray telescopes—the European XMM-Newton and NASA's Chandra satellite—astronomers discovered exactly this pattern of lines in three different directions in space. In each case, the WHIM was pooled up in a "wall" of galaxies, just as theory suggests.

The atoms of gas confined in these walls are, on average, around 50 centimeters apart. Even so, they cover such vast tracts of space that the clouds of WHIM may contain as much mass as all the visible galaxies in the universe put together.

The filamentary walls of galaxies and WHIM surround huge voids in the cosmos, where matter is far more tenuous: the voids are typically 50 million light years across. But even here we don't reach a true vacuum. Astronomers have found a smattering of galaxies within the voids, around one-tenth the average density of galaxies in the universe as a whole.

In our quest for the ultimate vacuum, we have literally come as far as we can. Assuming the interstellar gas in the voids is in rough proportion to the galaxies here, then the most tenuous part of our cosmos still contains some atoms. The spacing between them is around 5 meters.

So here, at the ends of the universe, lies the physicists' dream vacuum. Absolute emptiness. Take an experimental chamber the size of your living room to one of these voids and open the doors. When all the air has gone, there'll be nothing at all inside—or possibly just a single atom, if you're very unlucky.

To read more about vacuums, go to "The turbulent life of empty space" on page 126.

Zero, zip, zilch

We have heard that when it arrived in Europe, zero was treated with suspicion. We don't think of the absence of sound as a type of sound, so why should the absence of numbers be a number, argued its detractors. It took centuries for zero to gain acceptance. It is certainly not like other numbers. To work with it requires some tough intellectual contortions, as mathematician Ian Stewart explains.

Nothing is more interesting than nothing, nothing is more puzzling than nothing, and nothing is more important than nothing. For mathematicians, nothing is one of their favorite topics, a veritable Pandora's box of curiosities and paradoxes. What lies at the heart of mathematics? You guessed it: nothing.

Word games like this are almost irresistible when you talk about nothing, but in the case of math this is cheating slightly. What lies at the heart of math is related to nothing, but isn't quite the same thing. "Nothing" is—well, nothing. A void. Total absence of thingness. Zero, however, is definitely a thing. It is a number. It is, in fact, the number you get when you count your oranges and you haven't got any. And zero has caused mathematicians more heartache, and given them more joy, than any other number.

Zero, as a symbol, is part of the wonderful invention of "place notation." Early notations for numbers were weird and wonderful, a good example being Roman numerals, in which the number 1,998 comes out as MCMXCVIII—one thousand (M) plus one hundred less than a thousand (CM) plus ten less than a hundred (XC) plus five (V) plus one plus one plus one (III). Try doing arithmetic with that lot. So the symbols were used to record numbers, while calculations were done using the abacus, piling up stones in rows in the sand or moving beads on wires.

At some point, somebody got the bright idea of

representing the state of a row of beads by a symbol—not our current 1, 2, 3, 4, 5, 6, 7, 8, 9, but something fairly similar. The symbol 9 would represent nine beads in any row—nine thousands, nine hundreds, nine tens, nine units. The symbol's shape didn't tell you which, any more than the number of beads on a wire of the abacus did. The distinction was found in the position of the symbol, which corresponded to the position of the wire. In the notation 1,998, for instance, the first 9 means nine hundred and the second ninety.

The idea of place notation made it rather important to have a symbol for an empty row of beads. Without it, you couldn't tell the difference between 14, 104, 140 and 1,400. So in the beginning the symbol for zero was intimately associated with the concept of emptiness, rather than being a number in its own right. But by the 7th century, that had started to change. The Indian astronomer Brahmagupta explained that multiplying a number by 0 produced 0 and that subtracting 0 from a number left the number intact. By using 0 in arithmetic on the same footing as the other numbers, he showed that 0 had genuine numberhood.

Pandora's box was now wide open, and what burst forth was—nothing. And what a glorious, untamed, infuriating nothing it was.

The results obtained by doing arithmetic with zero were often curious, so curious sometimes that they had to be forbidden. Addition had the same effect as subtraction: the number stayed the same. Linguistic purists may object that leaving something unchanged hardly amounts to addition, but mathematicians generally prefer convenience to linguistic purity. Multiplication by zero, as Brahmagupta said, always yielded zero. It was with division that the serious trouble set in.

Dividing 0 by a non-zero number is easy: the result is 0. Why? Because 0 divided by 7, say, should be "whatever number gives 0 when multiplied by 7," and 0 is the only thing that fits the bill. But what is 1 divided by 0? It must be "whatever number gives 1 when multiplied by 0." Unfortunately, any number multiplied by 0 gives 0 not 1, so there's no such number. Division by zero is therefore forbidden, which is why calculators put up an error message if you try it.

Instead of forbidding fractions like 1 divided by 0, it is possible to release yet another irritant from Pandora's mathematical box—by defining 1 divided by 0 to be "infinity." Infinity is even weirder than zero; its use should always be accompanied by a government warning: "Infinity can seriously damage your calculations." Whatever infinity may be, it isn't a number in the usual sense. So mostly it's best to avoid things like 1 divided by 0.

Sorry: Pandora's curse is not so easily evaded. What about 0 divided by 0? Now the problem is not an absence of suitable candidates, but an embarrassment of them. Again, 0 divided by 0 should mean "whatever number gives 0 when multiplied by 0." But since this is true whatever number you use to divide 0 by, unless you're very careful, you can fall into many logical traps—the simplest such being the "proof" that 1 = 2 because both equal 0 when they are divided by 0. So 0 divided by 0 is also forbidden.

Alas, 0 divided by 0 was too seductive to stay forbidden for long. It is at the heart of calculus, the independent invention of Gottfried Wilhelm von Leibniz and Isaac Newton. Calculus was an extraordinary intellectual revolution, perhaps without historical parallel, because it gave us the idea that nature is at root mathematical.

In what sense is calculus about 0 divided by 0? Well, the underlying feature of calculus is the rate of change of some variable—how rapidly it is changing at a given instant. Here a formula or two seems unavoidable. Suppose some quantity x varies with time t, and write x(t) for its value at time t. This x might be how far your bike has traveled, so x(12 noon) = the Pig and Whistle pub. At that point your bike is probably not moving, unless an unfriendly local is stealing it, so the rate of change of x at 12 noon is zero. However, by some time t a bit after 2 PM, you are pedaling along the leafy byways at position x(t). How fast is your position changing, at that precise instant?

Newton's answer was to let time increase by a tiny amount—let's call it d. As t changes to t + d, your bike moves from x(t) to x(t + d)—say from level with a dozing sheep's left nostril to level with its right nostril. The amount by which your position changes is x(t + d)–x(t), the inter-nostril distance, and since it took you time d to achieve that change, the rate of change is (x(t + d) – x(t))/d; distance traveled divided by time taken to do so.

So far so good, but this expression represents the average rate of change over the time interval from t to t + d, not the rate of change at time t itself. However small d may be, even if it's 0.00000000001, this approach still doesn't give you the instantaneous rate of change. Newton's idea was to find the average rate of change over a time interval of length d, let d become zero, and see what you get.

In practice this leads to entirely sensible answers, but the procedure is mysterious. Enter Bishop Berkeley, best known for his philosophical writings on the problem of existence. Berkeley annoyed all the mathematicians by pointing out—correctly—that Newton's procedure amounts to dividing 0 by 0. Over a time interval of zero, your bike moves a distance of zero, and you're dividing one by the other.

Berkeley had an ulterior motive: he was upset by criticisms that religious faith was illogical, and he hit back by pointing out that calculus is illogical too. He did so in 1743 in a pamphlet entitled *The Analyst, Or a Discourse Addressed to an Infidel Mathematician Wherein it is examined whether the Object, Principles, and Inferences of the Modern Analysis are more distinctly conceived, or more evidently deduced, than Religious Mysteries and Points of Faith*. It contained the following: "First cast out the beam in thine own Eye; and then shalt thou see clearly to cast out the mote out of thy brother's Eye." Clearly the good bishop was a bit peeved; equally clearly, he did his homework on the math.

Newton tried to justify his calculations by appealing to physical intuition, and also by a rather weaselly explanation of how the method avoids dividing by zero. First you write down your equation using the variable d. The fraction involves dividing by d, but that's all right because at this stage you're saying that d is not zero. You then simplify your fraction until the d in the denominator disappears. Only then do you let d equal zero to get your answer.

How d can sometimes be allowed to be zero and sometimes not, Newton never really explained. Leibniz made a more mystical appeal to the "spirit of finesse" as opposed to the "spirit of logic" (which loosely translates as "I don't know what I'm doing, but hey, it works"). Berkeley claimed the method worked because of compensating errors, but missed the key point: why do the errors compensate?

In the end, the whole problem was tidied up by Karl Weierstrass, about 120 years later, who defined the elusive concept of a "limit." Rather than saying that d sometimes can and sometimes cannot be zero, you're actually calculating the value that the fraction approaches as d gets closer and closer to zero. And it works. So Newton and Leibniz created a new way of thinking about the world, while Berkeley's criticism, though right, was uncreative. The whole dispute, in fact, turned out to be about nothing.

Zero tolerance

Zero is a born troublemaker. Once mathematicians decided to treat it as a number, all the standard formulas had to be extended to embrace zero—with results that were not always intuitive. The most familiar example is powers. Take the fourth power of 5. This is 5^4 or $5 \times 5 \times 5 \times 5$. So clearly 5^0 must be five to the power of zero, or no fives multiplied together.

This is obviously not the way to think of it. Instead, what mathematicians do is to decide on some property of powers that they want to remain true. For instance, if you multiply powers together, the exponents add up: $5^2 \times 5^3 = (5 \times 5) \times (5 \times 5 \times 5) = 5^5$.

If you want 5^0 to be any use, it makes sense to retain this property, so that $5^0 \times 5^2$ must equal $5^{0+2} = 5^2 = 25$. That is, 5^0

× 25 equals 25. Therefore 5^0 must equal 1. For this reason, the standard convention is that the zeroth power of any number is 1—with one exception: 0^0. The above argument requires $0^0 \times 0$ to equal 0, sure—but now you can't divide out the 0s to conclude that $0^0 = 1$. In fact, just like 0 divided by 0, you have to deem 0^0 meaningless.

The same kind of approach determines the convention for zero factorial. Factorials, symbolized by an exclamation mark, are normally defined like this: $5! = 5 \times 4 \times 3 \times 2 \times 1$. Starting with the chosen number, reduce it one step at a time until you hit 1, then multiply the resulting numbers together.

But this doesn't help with 0!, since you have to stop before you start. The usual interpretation of n! is "the number of ways to arrange n things in order," but this also doesn't help, since it's not at all clear how many ways there are to arrange no things in order. The most plausible answer would seem to be "none, because there aren't any things to arrange," but that approach turns out to be misleading. Mathematicians prefer to preserve a general property of factorials, the pattern

$4! = 4 \times 3!$

$3! = 3 \times 2!$

$2! = 2 \times 1!$ and extend it to

$1! = 1 \times 0!$ Since $1! = 1$, this leads to the conclusion that $0! = 1$. And this is the convention that mathematicians employ.

To read more about zero, go to "Nothing in common" on page 158.

4

Surprises

Sherlock Holmes put it like this: "When you have eliminated the impossible, whatever remains, however improbable, must be the truth." Science has delivered more than its fair share of improbable, not to say mind-boggling explanations. Who would have guessed that something as seemingly simple as a vacuum would turn out not to be empty at all, that words really can kill and that the secret to a robust understanding of numbers is the absence of them?

The turbulent life of empty space

We've traced the vacuum from its discovery to its use as a store for antimatter and charted vacuums though the cosmos. The voids in these cases are "classical" vacuums—spaces that are empty-ish. Yet when scientists explored the nature of the vacuum at the smallest scale, they found a space seething with activity. Physicist Paul Davies introduces the bizarre world of the quantum vacuum.

"Nature abhors a vacuum." This sentiment, which first popped up in Greek philosophy some 2500 years ago, continues to excite debate among scientists and philosophers. The concept of a true void, apart from inducing a queasy feeling, strikes many people as preposterous or even meaningless. If two bodies are separated by nothing, should they not be in contact? How can "emptiness" keep things apart, or have properties such as size or boundaries?

While we continue to struggle with such notions, our idea of the vacuum has moved on. Empty space is richer than a mere absence of things—and it plays an indispensable part in much of modern physics.

Even among the ancient Greeks, the void divided loyalties. One influential line of thought, first apparent in the work of the philosopher Parmenides in the 5th century BC and today most commonly associated with Aristotle, held that empty space is really filled with an invisible medium. Proponents of the rival atomic theory, among them Leucippus and Democritus, disagreed. In their view, the cosmos consisted of a limitless void populated by tiny indestructible particles, or atoms, that came together in various combinations to form material objects.

Such metaphysical debates remained standard fare among philosophers until after the Middle Ages. The rise of modern science in the 17th century did little to settle them. Isaac Newton, like Aristotle, thought that the space between bodies must be filled with a medium, albeit of an unusual sort. It must be invisible, but also frictionless, as Earth ploughs through it on its way round the sun without meeting resistance.

Newton appealed to this substance as a reference frame for his laws of motion. They predicted, for example, that a spinning planet such as Earth would experience a centrifugal force that would make it bulge at the equator. This effect provided physical proof of the body's rotation, yet such a rotation, and thus the existence of a force, only made sense if there were some absolute frame of rest, a stationary viewpoint against which to compare the motion. This, said Newton, was the invisible medium that filled space.

Newton's German rival Gottfried Leibniz disagreed. He maintained that all motion, including rotation, was only to be judged relative to other bodies in the universe—the distant stars, for example. An observer on a merry-go-round in deep space would see the stars going round and at the same time feel a centrifugal force. According to Leibniz, if the stars were to vanish, so would the force; there was no need for an invisible medium in between.

Leibniz's position was argued forcefully in the 19th century by the German engineer and philosopher Ernst Mach, he of the Mach numbers used to denote aircraft speed. Mach proposed that centrifugal forces and related mechanical effects are caused by the gravitational action of distant matter in the universe. Albert Einstein was strongly influenced by Mach's ideas in formulating his theory of relativity, and was disappointed to find that Mach's principle did not emerge from it. In Einstein's theory, for example, a spinning black hole is predicted to have a bulging waist even when no other object exists.

During the 19th century, the nature of empty space began to occupy the thoughts of physicists in a new context: the mystery of how one charged body feels the pull of another, or how two magnets sense each other's presence. The chemist and physicist Michael Faraday's explanation was that charged or magnetic bodies created regions of influence—fields—around them, which other bodies experienced as a force.

But what, exactly, are these fields? One way physicists of the time liked to explain them was by invoking an invisible medium filling all of space, just as Newton had. Electric and magnetic fields can be described as strains in this medium, like those introduced to a block of rubber when you twist it. The medium became known as the luminiferous aether, or just the ether, and it had an enormous influence on 19th-century science. It was also popular with spiritualists, who liked its ghostliness, and invented obscure notions of "etheric bodies" said to survive death. When James Clerk Maxwell developed his theory unifying electricity and magnetism in the 1860s, the ether provided a natural habitat for the ghostly electromagnetic waves his theory predicted—things like radio waves and light.

So far, so good. Soon after Maxwell published his theory, however, the old problem of relative motion resurfaced. Even if our planet feels no friction as it slides through the ether, any movement relative to it should still produce measurable effects. Most notably, the speed of light should depend on the speed and direction of Earth's motion. Attempts to detect this experimentally by comparing the speed of light beams traveling in different directions failed to find any effect.

Einstein came to the rescue. His special theory of

relativity, published in 1905, suggests that a body's motion must always be judged relative to another body, and never to space itself, or space-filling invisible stuff. Electric and magnetic fields exist, but no longer as strains in any space-filling medium. Their strength and direction, and the forces they exert, change with the motion of the observer such that the speed of light is always measured to be the same, independently of how the observer moves. The ether became an unnecessary complication. While it is true to say that a region of space pervaded by an electric or magnetic field is not empty, the will-o'-the-wisp "stuff" it contains is a far cry from what we normally think of as matter. Fields might possess energy and exert pressure, but they are not made up of anything more substantial.

A decade or so on, however, a new twist cast the problem of empty space in a different light. It emerged from the theory of quantum mechanics. At the level of atoms, the clockwork predictability of the classical Newtonian universe broke down, to be replaced by a strange alternative set of rules. A particle such as an electron, for example, does not move from A to B along a precisely defined trajectory. At any given moment its position and motion will be, to a degree, uncertain.

What's true for an electron is true for all physical entities, including fields. An electric field, for instance, fluctuates in intensity and direction as a result of quantum uncertainty, even if the field is zero overall. Imagine a box containing no electric charges—in fact containing nothing but a vacuum—and made of metal so that no electric field can penetrate from the outside. According to quantum mechanics, there will still be an irreducible electric field inside the box, surging sometimes this way, sometimes that. Overall, these fluctuations average out to zero, so a crude measurement may not detect any electrical activity. A careful atomic-level measurement, on the other hand, will.

We now encounter an important point. Although the field strength of the fluctuations averages to zero, the energy does not, because an electric field's energy is independent of its direction. So how much energy resides in an empty box of a given size? Quick calculations on the basis of quantum theory lead to an apparently nonsensical conclusion: there is no limit. The vacuum is not empty. In fact, it contains an infinite amount of energy.

Physicists have found a way around this conundrum, but only by asking a different question. If you have two metal boxes of different size or shape, what is the difference in their quantum vacuum energy? The answer, it turns out, is tiny. But not so tiny that the difference cannot be measured in the lab, proving once and for all that the quantum fluctuations are real, and not just a crazy theoretical prediction.

So the modern conception of the vacuum is one of a seething ferment of quantum-field activity, with waves surging randomly this way and that. In quantum mechanics, waves also have characteristics of particles, so the quantum vacuum is often depicted as a sea of short-lived particles—photons for the electromagnetic field, gravitons for the gravitational field, and so on—popping out of nowhere and disappearing again. Wave or particle, what one gets is a picture of the vacuum that is reminiscent, in some respects, of the ether. It does not provide a special frame of rest against which bodies may be said to move, but it does fill all of space and have measurable physical properties such as energy density and pressure.

One of the most studied aspects of the quantum vacuum is its gravitational action. Out there in the cosmos there is a lot of space, all of it presumably chock-full of quantum-vacuum fluctuations. All those particles popping in and out of existence must weigh something. Perhaps that mass is enough to contribute to the total gravitating power of the universe; perhaps, indeed, enough to overwhelm the gravity of ordinary matter.

Finding the answer is a demanding task. We must account not just for electromagnetic fields, but all fields in nature—and we cannot be sure we have all of these pinned down yet. One general result can be readily deduced, however. In the event that the pressure of the quantum vacuum is negative (a negative pressure is a tension), the gravitational effect is also negative. That is, negative-pressure quantum-vacuum fluctuations serve to create a repulsive, or anti-gravitating, force.

Einstein had predicted that empty space would have such an anti-gravitational effect in 1917, before quantum mechanics. He couldn't put a number on the strength of the force, though, and later abandoned the idea. But it never completely went away. Back-of-the-envelope calculations today suggest that the quantum-vacuum pressure should indeed be negative in a space that has the geometry of our universe.

Sure enough, in the mid-1990s evidence began to accumulate from observations of far-off supernovae that a huge anti-gravitational force is causing the entire universe to expand faster and faster. The invisible quantum vacuum "ether" that is presumably at least partially responsible has recently been renamed "dark energy." In 2011, the observational work that led to this discovery earned astrophysicists Saul Perlmutter, Adam Riess and Brian Schmidt the Nobel prize in physics.

While quantum mechanics gives us a way to begin the calculation, a proper understanding of dark energy's strength and properties will probably require new physics, perhaps coming from string theory or some other attempt to bring all the fundamental forces of nature—including gravity, the perennial outsider—under one umbrella.

One thing is clear, however. The notion that space is a mere void with no physical properties is no longer tenable. Nature may abhor an absolute vacuum, but it embraces the quantum vacuum with relish. This is no semantic quibble. Depending on how dark energy works, the universe may continue on a runaway expansion or it might crush in on itself. The fate of the universe, it seems, lies in the properties of the vacuum.

To carry on reading about the vacuum, go to "Vacuum packed" on page 149.

When mind attacks body

As if the placebo effect were not strange enough, it also has an "evil twin": if the listener believes them, a few words can kill as surely as poison. Helen Pilcher finds out what this means for modern-day doctors, and for us as patients.

Late one night in a small Alabama cemetery, Vance Vanders had a run-in with the local witch doctor, who wafted a bottle of unpleasant-smelling liquid in front of his face and told him he was about to die and that no one could save him.

Back home, Vanders took to his bed and began to deteriorate. Some weeks later, emaciated and near death, he was admitted to the local hospital, where doctors were unable to find a cause for his symptoms or slow his decline. Only then did his wife tell one of the doctors, Drayton Doherty, of the hex.

Doherty thought long and hard. The next morning, he called Vanders's family to his bedside. He told them that the previous night he had lured the witch doctor back to the cemetery, where he had choked him against a tree until he explained how the curse worked. The medicine man had, he said, rubbed lizard eggs into Vanders's stomach, which had hatched inside his body. One reptile remained, which was eating Vanders from the inside out.

Doherty then summoned a nurse who had, by prior arrangement, filled a large syringe with a powerful emetic. With great ceremony, he inspected the instrument and injected its contents into Vanders's arm. A few minutes later, Vanders began to gag and vomit uncontrollably. In the midst of it all, unnoticed by everyone in the room, Doherty produced his pièce de résistance—a green lizard he had stashed in his black bag. "Look what has come out of you, Vance!" he cried. "The voodoo curse is lifted."

Vanders did a double take, fell back on the bed, and then drifted into a deep sleep. When he woke the next day, he was alert and ravenous. He quickly regained his strength and was discharged a week later.

The facts of this case, which happened in 1938, were corroborated by four medical professionals. Perhaps the most remarkable thing about it is that Vanders survived, as there are numerous documented instances from many parts of the globe of other people dying after being cursed.

With no medical records and no autopsy results, there's no way to be sure exactly how these people met their end. The common thread in these cases, however, is that a respected figure puts a curse on someone, perhaps by chanting or pointing a bone at them. Soon afterward, the victim dies, apparently of natural causes.

You might think this sort of thing is increasingly rare, and limited to remote tribes. But according to Clifton Meador, a doctor at Vanderbilt School of Medicine in Nashville, Tennessee, who has documented similar cases, the curse has taken on a new form.[1]

Take Sam Shoeman, who was diagnosed with end-stage liver cancer in the 1970s and given just months to live. Shoeman duly died in the allotted time frame—yet the autopsy revealed that his doctors had got it wrong. The tumor was tiny and had not spread. "He didn't die from cancer, but from believing he was dying of cancer," says Meador. "If everyone treats you as if you are dying, you buy into it. Everything in your whole being becomes about dying."

Cases such as Shoeman's may be extreme examples of a far more widespread phenomenon. Many patients who suffer harmful side effects, for instance, may do so only because they have been told to expect them. What's more, people who believe they have a high risk of certain diseases are more likely to get them than people with the same risk factors who believe they have a low risk. It seems modern witch doctors wear white coats and carry stethoscopes.

The idea that believing you are ill can make you ill may seem far-fetched, yet rigorous trials have established beyond doubt that the converse is true—that the power of suggestion can improve health. This is the well-known placebo effect. Placebos cannot produce miracles, but they do produce measurable physical effects.

The placebo effect has an evil twin: the nocebo effect, in which dummy pills and negative expectations can produce harmful effects. The term "nocebo," which means "I will harm," was not coined until the 1960s, and the phenomenon has been far less studied than the placebo effect. It's not easy, after all, to get ethical approval for studies designed to make people feel worse.

What we do know suggests the impact of nocebo is far-reaching. "Voodoo death, if it exists, may represent an extreme form of the nocebo phenomenon," says anthropologist Robert Hahn of the US Centers for Disease Control and Prevention in Atlanta, Georgia, who has studied the nocebo effect.

In clinical trials, around a quarter of patients in control groups—those given supposedly inert therapies—experience negative side effects.[2]

The severity of these side effects sometimes matches those associated with real drugs. A retrospective study of 15 trials involving thousands of patients prescribed either beta blockers or a control showed that both groups reported comparable levels of side effects, including fatigue, depressive symptoms and sexual dysfunction. A similar number had to withdraw from the studies because of them.

Occasionally, the effects can be life-threatening (see box, "The overdose"). "Beliefs and expectations are not only conscious, logical phenomena, they also have physical consequences," says Hahn.

The overdose

Depressed after splitting up with his girlfriend, Derek Adams took all his pills . . . then regretted it. Fearing he might die, he asked a neighbor to take him to the hospital, where he collapsed. Shaky, pale and drowsy, his blood pressure dropped and his breaths came quickly.

Bafflingly, lab tests and toxicology screening came back clear. Over the next four hours Adams received six liters of saline, but improved little.

Then a doctor arrived from the clinical trial of an antidepressant in which Adams had been taking part. Adams had enrolled in the study about a month earlier. Initially he had felt his mood buoyed, but an argument with his ex-girlfriend saw him swallow the 29 remaining tablets.

The doctor revealed that Adams was in the control group. The pills he had "overdosed" on were harmless. Hearing this, Adams was surprised and tearfully relieved. Within 15 minutes he was fully alert, and his blood pressure and heart rate had returned to normal.

Nocebo effects are also seen in normal medical practice. Around 60 percent of patients undergoing chemotherapy start feeling sick before their treatment. "It can happen days before, or on the journey on the way in," says clinical psychologist Guy Montgomery from Mount Sinai School of Medicine in New York. Sometimes the mere thought of treatment or the doctor's voice is enough to make patients feel unwell. This "anticipatory nausea" may be partly due to conditioning—when patients subconsciously link some part of their experience with nausea—and partly due to expectation.

Alarmingly, the nocebo effect can even be contagious. Cases where symptoms without an identifiable cause spread through groups of people have been around for centuries, a phenomenon known as mass psychogenic illness.[3] One outbreak (see box, "It's catching") inspired a study by psychologists Irving Kirsch and Giuliana Mazzoni of the University of Hull.[4]

They asked some of a group of students to inhale a sample of normal air, which all participants were told contained "a suspected environmental toxin" linked to headache, nausea, itchy skin and drowsiness. Half of the participants also watched a woman inhale the sample and apparently develop these symptoms. Students who inhaled were more likely to report these symptoms than those who did not. Symptoms were also more pronounced in women, particularly those who had seen another apparently become ill—a bias also seen in mass psychogenic illness.

The study shows that if you hear of or observe a possible side effect, you are more likely to develop it yourself.

It's catching

In November 1998, a teacher at a Tennessee high school noticed a "gasoline-like" smell, and began complaining of headache, nausea, shortness of breath and dizziness. The school was evacuated and over the next week more than 100 staff and students were admitted to the local emergency room complaining of similar symptoms.

After extensive tests, no medical explanation for the reported illnesses could be found. A questionnaire a month later revealed that the people who reported symptoms were more likely to be female, and to have known or seen a classmate who was ill. It was the nocebo effect on a grand scale, says psychologist Irving Kirsch at the University of Hull. "There was, as far as we can tell, no environmental toxin, but people began to feel ill."

Kirsch thinks that seeing a classmate develop symptoms shaped expectancies of illness in other children, triggering mass psychogenic illness. Outbreaks occur all over the world. In Jordan in 1998, 800 children apparently suffered side effects after a vaccination and 122 were admitted to hospital, but no problem was found with the vaccine.[7]

That puts doctors in a tricky situation. "On the one hand people have the right to be informed about what to expect, but this makes it more likely they will experience these effects," says Mazzoni.

This means doctors need to choose their words carefully so as to minimize negative expectations, says Montgomery. "It's all about how you say it."

Hypnosis might also help. "Hypnosis changes expectancies, which decreases anxiety and stress, which improves the outcome," Montgomery says. "I think hypnosis could be applied to a wide variety of symptoms where expectancy plays a role."

Is the scale of the nocebo problem serious enough to justify such countermeasures? We just don't know, because so many questions remain unanswered. In what circumstances do nocebo effects occur? And how long do the symptoms last?

It appears that, as with the placebo response, nocebo effects vary widely, and may depend heavily on context. Placebo effects in clinical settings are often much more potent than those induced in the laboratory, says Paul Enck, a psychologist at the University Hospital in Tübingen, which suggests the nocebo problem may have profound effects in the real world. For obvious reasons, though, lab experiments are designed to induce only mild and temporary nocebo symptoms.

It is also unclear who is susceptible. A person's optimism or pessimism may play a role, but there are no consistent personality predictors. Both sexes can succumb to mass psychogenic illness, though women report more symptoms than men. Enck has shown that in men, expectancy rather than conditioning is more likely to influence nocebo symptoms. For women, the opposite is true. "Women tend to operate more on past experiences, whereas men seem more reluctant to take history into a situation," he says.

What is becoming clear is that these apparently psychological phenomena have very real consequences in the brain. In 2008, using PET scans to peer into the brains of people given a placebo or nocebo, Jon-Kar Zubieta of the University of Michigan, Ann Arbor, found that nocebo effects were linked with a decrease in dopamine and opioid activity.[5] This would explain how nocebos can increase pain. Placebos, unsurprisingly, produced the opposite response.

Meanwhile, Fabrizio Benedetti of the University of Turin Medical School has found that nocebo-induced pain can be suppressed by a drug called proglumide, which blocks receptors for a hormone called cholecystokinin (CCK). Normally, expectations of pain induce anxiety, which activates CCK receptors, enhancing pain.

The ultimate cause of the nocebo effect, however, is not neurochemistry but belief. According to Hahn, surgeons are often wary of operating on people who think they will die—because such patients often do. And the mere belief that one is susceptible to a heart attack is itself a risk factor. One study found that women who believe they are particularly prone to heart attack are nearly four times as likely to die from coronary conditions as other women with the same risk factors.[6]

Despite the growing evidence that the nocebo effect is all too real, it is hard in this rational age to accept that people's beliefs can kill them. After all, most of us would laugh if a strangely attired man leapt about waving a bone and told us we were going to die. But imagine how you would feel if you were told the same thing by a smartly dressed doctor with a wallful of medical degrees and a computerful of your scans and test results. The social and cultural background is crucial, says Enck.

Meador argues that Shoeman's misdiagnosis of cancer and subsequent death shares many of the crucial elements found in hex death. A powerful doctor pronounces a death sentence, which is accepted unquestioningly by the "victim" and his family, who then start to act upon that belief. Shoeman, his family and his doctors all believed he was dying from cancer. It became a self-fulfilling prophecy.

"Bad news promotes bad physiology. I think you can persuade people that they're going to die and have it happen," Meador says. "I don't think there's anything mystical about it. We're uncomfortable with the idea that words or symbolic actions can cause death because it challenges our biomolecular model of the world."

Perhaps when the biomedical basis of voodoo death is revealed in detail we will find it easier to accept that it is real—and that it can affect any one of us.

Ride the celestial subway

It's not just humans who sometimes "feel nothing." There are places in the solar system where gravity falls to zero. Here, spacecraft, asteroids and other interplanetary detritus "feel nothing"—no gravitational attraction in any direction. Without their own propulsion, they would stay marooned forever. Mathematician Ian Stewart describes how we can use these points as interchanges on the Interplanetary Superhighway.

What is the most efficient route from the Earth to the moon or the planets? According to NASA engineers, the answer is simple. Travel the way Londoners do: go by tube.

The idea is not entirely original. Peter Hamilton's science-fiction novel *Pandora's Star* portrays a future in which people travel by train to planets encircling distant stars. True, the railway lines have to pass through a wormhole—a shortcut through space-time—but once you can build wormholes to order, it's entirely logical to use trains. Much earlier, E. E. "Doc" Smith came up with the hyperspatial tube, used by malevolent aliens to invade human worlds from another dimension. And C. C. MacApp imagined a universe in which star systems were connected by a system of tubes, through which spaceships could travel faster than light.

Although we don't yet have wormholes, extra dimensions or faster-than-light tunnels through space, mathematicians have discovered that our solar system does possess something remarkably similar to the wild inventions of these authors' imaginations: a network of tubes perfectly suited to space travel. The tubes can be seen only through mathematical eyes, because their walls are defined by the combined gravitational fields of all the bodies in the solar system, but they are real enough, for all that. If we could visualize the ever-changing fields that control how the planets, moons, asteroids and comets move, we would be able to see a network of tubes that swirl and twirl along with the planets as they perform their endless gravitational dance. NASA's engineers already refer to the network as the "Interplanetary Superhighway": science fiction wins again.

The existence of this interplanetary network would have surprised the pioneers of celestial mechanics. Yet, paradoxically, it was those pioneering efforts centuries ago that provided the first hint of the tubes' presence. Long ago, the planets were seen as unruly members of the celestial club. The fixed stars appeared to be bolted to the heavenly sphere, spinning once every day about the poles. The planets, though, were wanderers—in Greek, *planan* means "to lead astray." Their place in the heavens, relative to the stars, was not fixed, and their motions were complex and difficult to describe. They seemed to move of their own volition, which is perhaps why they were identified with gods.

Johannes Kepler and Isaac Newton taught us that such appearances are deceptive: the planets are no more unruly than the stars. Planets revolve around the sun in tidy elliptical orbits, and their apparent wild gyrations are the result of a combination of their movements with those of Earth as it follows a similarly elliptical path. The solar system came to be seen as a well-oiled piece of celestial clockwork, as if invisible gears and cogs were turning the great wheels of the planetary system. The devil was in the detail, of course, and the orbits were not precisely elliptical, because each world affected the movement of the others. But it could all be summed up in Newton's law of gravity: every body in the universe attracts every other body with a force that is proportional to their masses and inversely proportional to the square of the distance between them.

Since the law of gravity is so simple, it was only natural to imagine that the movements of planets and moons must be simple too. The solar system's celestial dance was thought to be slow and stately, heavily constrained by natural law. While the end result might be complicated, it could never be surprising.

But that is just not true. Consider, for example, the unruly comet Oterma. A century ago, Oterma's orbit was well outside the orbit of Jupiter until, after a close encounter with that giant planet, its orbit shifted inside Jupiter's. After another close encounter, it switched back outside Jupiter. We can confidently predict that Oterma will continue to switch orbits in this way every few decades. If this all seems a far cry from Kepler's and Newton's tidy elliptical orbits, it is, and with good reason. The orbits predicted by Newtonian gravity are only elliptical when no other bodies exert a significant gravitational pull. In fact, the solar system is full of other bodies, and they can make a significant—and surprising—difference. Which is where those tubes come in.

The tubes are a feature of "gravitational topography." The solar system is like an alpine landscape—but with the gravitational fields of the sun, planets and their moons providing the mountains and hills. A gravitational contour map of the solar system has similar features to a terrestrial contour map. There are closely packed rings where the gravitational field strength peaks near the sun, say, and there are flat contourless "valley floors" where the gravitational fields of two neighboring bodies cancel out. And just as Victorian railway engineers realized they could run trains most easily along the contours of a landscape, mathematicians have realized that a spacecraft will run most efficiently along the gravitational contours of space.

There is a complication, however. The trajectory of a spacecraft is influenced by its own speed as well as the local gravitational fields. In the late 1960s, Richard McGehee of the University of Minnesota in Minneapolis and the late Charles Conley pointed out that each contour's path is effectively surrounded by a nested set of tubes, one inside the other. Each tube corresponds to a particular choice of speed: the further away it is from the optimal speed for following a particular path, the wider the tube becomes. A spacecraft can travel along one of these tubes, following a gravitational contour at a certain speed, without expending fuel. When it needs to change course, it can do so by applying a little power boost in the right direction to move onto another contour.

Better still, there is an even more efficient way to change course: use the Interplanetary Superhighway's natural interchanges. The calculations that disclosed the existence of these interchanges were completed more than 200 years ago by Joseph-Louis Lagrange. They revealed that in a system consisting of just two bodies—the Earth and moon, say—there are five places where the gravitational fields of the two bodies cancel out exactly (in the frame of reference rotating with the two bodies). Three are in line with both Earth and moon: L1 lies between them, L2 is on the far side of the moon, and L3 is on the far side of Earth. There are also the two "Trojan points" L4 and L5, in the same orbit as the moon but 60 degrees ahead of it or behind it. As the moon orbits the Earth, the Lagrange points orbit it too. Other pairs of bodies also have Lagrange points—Earth/sun, Jupiter/sun, Titan/Saturn, and so on. At some of the Lagrange points there also exist "halo orbits" in which a body can stably loop about the Lagrange point.

Now imagine the gravity landscape surrounding a spacecraft sitting at the Earth/moon L1 point. If the craft is given a small push, it will start to run "downhill," following a tube that leads into an orbit around either Earth or the moon. The good news for space engineers is that these tubes trace out the most energetically efficient path from Earth to the moon. To make the journey, you would first give a little kick to move out of Earth orbit into the tube that runs to L1. Once there, you can nudge your spacecraft into the tube from L1 to the moon and let gravity do the rest.

The beauty of all this is that the tubes slinking their way through the solar system can be interconnected. Oterma's orbit, for instance, follows two tubes that meet near Jupiter. One tube lies inside Jupiter's orbit, the other outside. Where they meet, the comet can switch tubes—or not, depending on rather subtle effects of Jovian and solar gravity. Once inside a tube, Oterma is stuck there until the tube returns it to the junction; Oterma has no propulsion so it can't choose its trajectory, and will always remain near Jupiter.

Spacecraft, however, can do almost what they like—and Jupiter is not the only junction. The way to plan an efficient mission profile, then, is to work out which tubes are relevant to your choice of destination. You then route your spacecraft along the tube to a Lagrange point, and when it gets there you give it a quick burst on the motors to redirect it to the next Lagrange point on the route . . . and so it goes on.

It gets better. The dynamics of a spacecraft near L1, for example, are chaotic, so you can achieve large changes to the trajectory through very small changes of position or speed. By exploiting chaos, spacecraft can be redirected to other destinations—again, in a very fuel-efficient, though possibly slow, manner. This trick was used in the mid-1980s to redirect the almost dead International Cometary Explorer to rendezvous with comet Giacobini-Zinner. It was used again for NASA's Genesis mission, which crash-landed in Utah in 2004 after capturing samples of the solar wind.

Though calculating and exploiting the topography of the energy landscape requires some clever mathematics, today's computers have made it almost routine. In 2000, the tube technique was used by Wang Sang Koon of Caltech, the late Jerrold Marsden, Shane Ross, now at Virginia Tech in Blacksburg, and Martin Lo of NASA's Jet Propulsion Laboratory in Pasadena, to find a "Petit Grand Tour" of the moons of Jupiter, ending with a capture orbit round Europa.[1] The path requires a gravitational boost near Ganymede followed by a tube trip to Europa. A more complex route, requiring even less energy, includes Callisto as well. And in 2005, Michael Dellnitz, Marcus Post and Bianca Thiere, then at the University of Paderborn in Germany, with Oliver Junge of Technical University of Munich, used tubes to plan an energy-efficient mission from Earth to Venus.[2] The main tube here links the sun/Earth L1 point to the sun/Venus L2 point. Its low-thrust engines would require only one-third of the fuel used by the European Space Agency's Venus Express mission; the price paid is a lengthening of the journey time from 150 days to about 650 days. Future interplanetary missions for which fuel rather than time is of the essence will no doubt make routine use of tubes.

Vacuum packed

As we've seen, part way through the 20th century, the newly created quantum theory proposed the outlandish idea that all space is bubbling with energy and short-lived particles. Ever eager to exploit a new phenomenon, scientists asked: "Can we get anything useful out of this quantum vacuum?" David Harris explores the attempts to squeeze something from nothing.

"Nothing will come of nothing." Shakespeare's epigram seems the kind of self-evident statement that only poets and philosophers would argue over. And physicists like Chris Wilson.

In 2011, Wilson and his team at the Chalmers University of Technology in Gothenburg provided what seems a particularly egregious case of something for nothing. They claimed to have conjured up light from nowhere simply by squeezing empty space.[1]

That would be the latest manifestation of a quantum quirk known as the Casimir effect: the notion that a perfect vacuum, the very definition of nothingness in the physical world, contains a latent power that can be harnessed to move objects and make stuff.

Sightings of this vacuum action have been mounting over the past decade or so, leading some physicists to propose a new generation of nanoscale machines to take advantage of it, and others even to suggest a leading role for vacuum energy in determining the origin and fate of the cosmos. Others remain to be convinced. So what's the true story?

The idea that a vacuum is a seething sea of something

can be traced back to the early decades of quantum physics. In the late 1920s, the German physicist Werner Heisenberg came up with his famous uncertainty principle, which says that some pairs of measurable quantities are intimately connected: the more you know about the one, the less you know about the other.

Energy and time make up one such pair. That means you cannot measure the energy of a physical system with perfect precision unless you take infinite time to perform your measurement. In reality, then, it follows that the energy of the vacuum can never be pinned down precisely.

According to quantum theory, even a perfect vacuum is filled with wave-like fields that fluctuate constantly, producing a legion of ephemeral particles that continually pop out of nowhere only to disappear again, filling the vacuum with what's called "zero-point energy."

This recasting of the vacuum gave fresh impetus to the centuries-old debate about the nature of nothingness. But evidence also began to accumulate that the newly lively vacuum had practical effects. Observe atoms carefully enough and you see a tiny effect known as Lamb shift, in which vacuum fluctuations jostle an orbiting electron, subtly altering its energy. Something similar can be invoked to explain how electrons sometimes spontaneously jump between two atomic energy states, giving off photons of light.

But the Dutch physicist Hendrik Casimir's suggestion was the most eye-catching. In 1948, Casimir, together with his colleague Dirk Polder, was trying to understand how colloids exist in a stable equilibrium. Colloids are mixtures in which one type of substance is dispersed through another, like fat globules in the watery solution of milk.

Forces between the molecules in such a medium drop off more quickly with distance than expected when the classical electromagnetic attractions and repulsions, called van der Waals forces, are taken into account. It is as if something is pulling the constituent molecules closer together, giving the mixture extra stability.

Following a tip-off from the Danish quantum doyen Niels Bohr, Casimir calculated that this something could be vacuum action. Working out the effects of vacuum fluctuations in a colloid's complex molecular brew was impossibly involved. So Casimir considered a simple model system of two parallel metallic plates, and showed that the fluctuations could produce just the right enhanced attraction between them. His explanation was that the two plates limit the wavelength of vacuum fluctuations in the space between. Outside those confines, the fluctuations can have any wavelength they choose. With more waves outside than in, a pressure pushes inward on the plates (see figure on page 155).

The effect is tiny: two plates 10 nanometers apart feel a force comparable to the gentle burden of the atmosphere on our heads. Such a minuscule contribution is easily disguised by a legion of other effects, such as residual electrostatic attractions between charges on the plates' surfaces. That makes confirming its existence extremely tough. "You need to know that you're really measuring the Casimir force," says experimentalist Hong Tang of Yale University. What's more, it is not easy to align plates to be perfectly parallel, while calculating the expected effect for other, more complex geometries takes some sophisticated mathematics.

It was only in 1996 that Steven Lamoreaux, a physicist then at the University of Washington in Seattle, made a breakthrough. Taking elaborate precautions to exclude all other effects, he found a tiny residual force pulling a metal plate and a spherical lens together.[2] The Casimir effect, it seemed, was not a theorist's pipe dream: vacuum action was a real effect.

Since then, a steady trickle of results has confirmed other long-standing theoretical predictions. Soviet physicist Evgeny Lifshitz proposed in 1955 that the size of vacuum fluctuations would grow with rising temperature, resulting in a force that is more potent over longer distances. In February 2011, Lamoreaux, now at Yale University, and his team confirmed that this is indeed the case.[3]

As for the work of Wilson's team, their results, published in November 2011, support a four-decade-old prediction that turns the logic of the original Casimir effect on its head. Rather than using the vacuum's pop-up particles to shift their surroundings, if you move a vacuum's surroundings fast enough, you can make real photons of light. In some quarters, this idea is controversial—but it is the most dramatic putative demonstration of the vacuum's powers to date (see box, "Light from speeding mirrors").

As sightings of such effects have multiplied, so have thoughts that we might harness them for our own devices. A popular proposal is to use the vacuum's energy to give nanoscale machines an additional kick. That requires something a little different from the original Casimir force, whose attractive effects are more likely to gum up

Light from speeding mirrors

In 1970, American physicist Gerald Moore proposed reversing the logic of the Casimir effect. He envisaged rapidly accelerating mirrors that would squeeze the vacuum fluctuations in the space between them so violently that they would give up some of their energy in the form of photons.[7]

In practice it is not possible to accelerate even a small macroscopic mirror fast enough to produce this "dynamical" Casimir effect, so in 2011 Chris Wilson and his team from the Chalmers University of Technology in Gothenburg used rapidly varying electrical currents to simulate the effect of mirrors accelerating to something like a quarter of the speed of light. The result was the simultaneous production of pairs of photons from the vacuum, exactly as Moore had predicted.[8]

Wilson thinks there could be some exciting applications. During the era of inflation thought to have taken place right after the big bang, the boundary of the universe itself would have expanded at near the speed of light, leading to the creation of photons through the dynamical Casimir effect. "It is rather difficult to create your own big bang in the lab," says Wilson. "Our set-up or a similar one might be used to simulate these effects, essentially doing table-top cosmology."

Just as the original Casimir effect is disputed, however (see main story), not everyone is convinced that this interpretation of the experiment is right. One physicist, who preferred not to be named, says that as nothing in the experiment actually moves, it does not demonstrate the dynamical Casimir effect at all. Instead, it is just another "solid and interesting" example of a well-known effect in which some of a quantum circuit's electrical energy is emitted as light. The mathematical description of the two effects is very similar, he says, but "one should never mistake mathematics for reality."

Since the preliminary version of their paper was circulated, Wilson's team has carried out additional tests that Wilson thinks defuse such criticisms, although he acknowledges there are still dissenting voices.

"We did a number of sanity checks ruling out various spurious effects that could have masqueraded as the effect, including showing that we were starting from the vacuum state," he says. "But for some people, the dynamical Casimir effect will never be anything but a literal moving mirror."

the components of any mini-machine—a phenomenon referred to as static friction or "stiction."

By tweaking the geometries or material properties of the structures used to confine the vacuum, however, it should be possible to reverse the direction of the Casimir effect, creating an outward pressure to push two objects apart. In 2008, Steven Johnson and his colleagues at the Massachusetts Institute of Technology calculated that by adding a series of interleaving metal brackets, zipper-style, to the faces of the two metal plates you could in theory make the net force between them repulsive. A more recent study by Stanislav Maslovski and Mário Silveirinha of the University of Coimbra, Portugal, has indicated a similar effect using nanoscale metallic rods to create areas of repulsive force that can levitate a nanoscale metal bar.[4]

These forces could help nanoscale components such as switches, gears, bearings or motor parts to operate without jamming. Putting such devices into practice might not be easy, though. For a start, it would require

Wave squeeze

In quantum theory, space is filled with wave-like fields containing energy. Between two plates these fluctuations are limited, leaving a new inward force (at top). More complex geometrics (center) can create a repulsive force. Move the plates together quickly, and the wavelengths are suddenly restricted, forcing the vacuum to give up energy as photons (bottom)

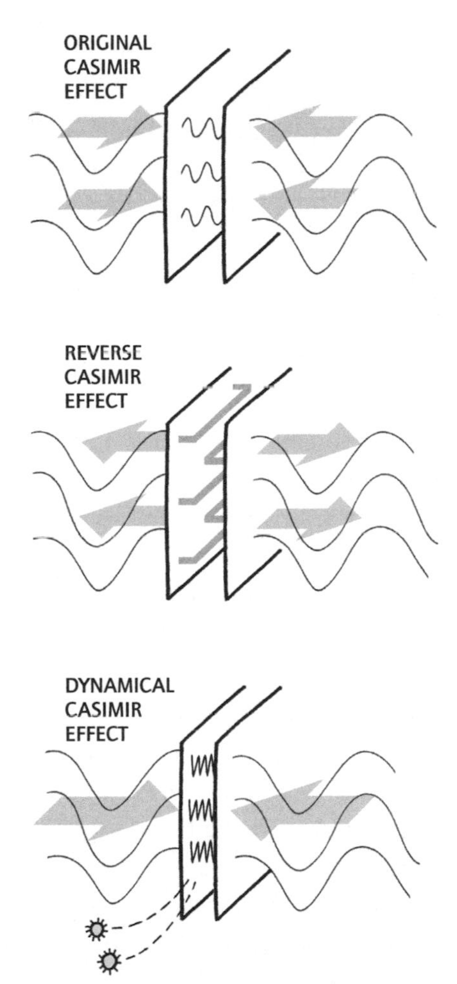

components with atomic-scale polishing: look on a small enough scale—a thousand atoms or so—and metal surfaces usually thought of as smooth have patchy, crystal-like structures that would confine vacuum fluctuations in different ways, affecting the size of the Casimir force. For moving objects, things become even trickier.

Such complications are surmountable: in 2009 Federico Capasso and his group at Harvard University measured what appeared to be repulsive Casimir forces in a gold cantilever suspended in bromobenzene liquid above a silicon surface.[5] The forces generated were mere tens of piconewtons—but when you are trying to move nanoscale particles, a piconewton goes a long way. Nevertheless, there are still hurdles to be overcome before Casimir devices are everyday reality, says Johnson. "It is an experimental question—can we make devices this small and sensitive?" he says. "And it is also a theoretical question of whether we can design interesting uses for the Casimir force once the experimental capabilities arrive."

There is a more fundamental objection, however. The litany of theoretical predictions gradually being turned into experimental reality invites a simple conclusion: vacuum fluctuations are real, and are responsible for what we call Casimir effects. But not all physicists buy that.

Their unease lies in calculations done by Casimir and Polder even before they settled on vacuum fluctuations as the explanation for the weakened van der Waals force. These showed that much the same weakening could be achieved simply by taking into account the finite time the force takes to be transmitted over large enough distances, such as between two plates separated by tens or hundreds of nanometers. That idea was revived and bolstered by calculations in the 1970s by the Nobel-prizewinning physicist Julian Schwinger. He never believed in the reality of vacuum fluctuations and developed a version of quantum field theory, which he called source theory, to do away with them. In this picture, the Casimir effect pops out just by taking into account the quantum interaction of charged matter, with no vacuum action at all.

Robert Jaffe, a particle theorist at the Massachusetts Institute of Technology, suggests the only reason the vacuum interpretation has gained such currency is because its mathematics happens to be a lot simpler. "There is a flippant way people refer to the Casimir effect as evidence for real vacuum fluctuations," he says. "But there is no evidence that the vacuum fluctuations exist in the absence of matter." Similarly, other effects invoked as proof of their reality—the Lamb shift and the spontaneous emission of photons from atoms—can be described purely as the result of charge interactions.

If this is so, it could have repercussions for more than our attempts to fine-tune the workings of nanomachines. The realization in the past couple of decades that the universe's expansion is accelerating—a phenomenon ascribed to a mysterious "dark energy"—has fueled a new interest in the power of the vacuum. At the moment, our best calculations of the vacuum's hidden energy come up with a figure some 120 orders of magnitude larger than the amount needed to bring about the cosmic acceleration, a mismatch that counts perhaps as the worst-ever prediction in physics. Yet observations of the Casimir effect are still eagerly seen as evidence for a power that might determine our cosmic fate.

Schwinger's original calculations were part of his wider attempt, ultimately unsuccessful, to banish vacuum fluctuations from quantum field theory. The truth may well lie uncomfortably in the middle: we may never be able to convince ourselves of the reality of vacuum energy, because any attempt to do so brings some form of matter into the equation. As philosophers of science Svend Rugh and Henrik Zinkernagel wrote in 2001, "It seems impossible to decide whether the effects result from the vacuum 'in itself'. . . or are generated by the introduction of the measurement arrangement."[6]

Wilson hopes that the photons emerging from his apparatus in Sweden, if confirmed by other groups, will provide the final illumination to prove the reality of vacuum fluctuations. Equally, as our ability to construct filigree nanomachines and so test the Casimir effect increases in coming years, perhaps some deviation from the predictions will give us a definitive handle on where the effects come from. Can nothing truly come of nothing? We might still have cause to speak again.

Nothing in common

Over the centuries zero has not only suffered trials and tribulations, it has also caused them. Mathematicians have persevered with it because its benefits massively outweigh its problems. Doing math without it today seems inconceivable. Here's mathematician Ian Stewart on one example of zero displaying its value.

The mathematicians' version of nothing is the empty set. This is a collection that doesn't actually contain anything, such as my own collection of vintage Rolls-Royces. The empty set may seem a bit feeble, but appearances deceive; it provides a vital building block for the whole of mathematics.

It all started in the late 1800s. While most mathematicians were busy adding a nice piece of furniture, a new room, even an entire storey to the growing mathematical edifice, a group of worrywarts started to fret about the cellar. Innovations like non-Euclidean geometry and Fourier analysis were all very well—but were the underpinnings sound? To prove they were, a basic idea needed sorting out that no one really understood. Numbers.

Sure, everyone knew how to do sums. Using numbers wasn't the problem. The big question was what they were. You can show someone two sheep, two coins, two albatrosses, two galaxies. But can you show them two?

The symbol "2"? That's a notation, not the number itself. Many cultures use a different symbol. The word "two"? No, for the same reason: in other languages it might be *deux* or *zwei* or *futatsu*. For thousands of years humans had been using numbers to great effect; suddenly a few deep thinkers realized no one had a clue what they were.

An answer emerged from two different lines of thought: mathematical logic, and Fourier analysis, in which a complex waveform is represented as a combination of simple sine waves. These two areas converged on one idea. Sets.

A set is a collection of mathematical objects—numbers,

shapes, functions, networks, whatever. It is defined by listing or characterising its members. "The set with members 2, 4, 6, 8" and "the set of even integers between 1 and 9" both define the same set, which can be written as {2, 4, 6, 8}.

Around 1880 the mathematician Georg Cantor developed an extensive theory of sets. He had been trying to sort out some technical issues in Fourier analysis related to discontinuities—places where the waveform makes sudden jumps. His answer involved the structure of the set of discontinuities. It wasn't the individual discontinuities that mattered, it was the whole class of discontinuities.

One thing led to another. Cantor devised a way to count how many members a set has, by matching it in a one-to-one fashion with a standard set. Suppose, for example, the set is {Doc, Grumpy, Happy, Sleepy, Bashful, Sneezy, Dopey}. To count them we chant "1, 2, 3 . . ." while working along the list: Doc (1), Grumpy (2), Happy (3), Sleepy (4), Bashful (5), Sneezy (6) Dopey (7). Right: seven dwarfs. We can do the same with the days of the week: Monday (1), Tuesday (2), Wednesday (3), Thursday (4), Friday (5), Saturday (6), Sunday (7).

Another mathematician of the time, Gottlob Frege, picked up on Cantor's ideas and thought they could solve the big philosophical problem of numbers. The way to define them, he believed, was through the deceptively simple process of counting.

What do we count? A collection of things—a set. How do we count it? By matching the things in the set with a standard set of known size. The next step was simple but devastating: throw away the numbers. You could use the dwarfs to count the days of the week. Just set up the correspondence: Monday (Doc), Tuesday (Grumpy) . . . Sunday (Dopey). There are Dopey days in the week. It's a perfectly reasonable alternative number system. It doesn't (yet) tell us what a number is, but it gives a way to define "same number." The number of days equals the number of dwarfs, not because both are seven, but because you can match days to dwarfs.

What, then, is a number? Mathematical logicians realized that to define the number 2, you need to construct a standard set which intuitively has two members. To define 3, use a standard set with three members, and so on. But which standard sets to use? They have to be unique, and their structure should correspond to the process of counting. This was where "the empty set" came in and solved the whole thing by itself.

Zero is a number, the basis of our entire number system. So it ought to count the members of a set. Which set? Well, it has to be a set with no members. These aren't hard to think of: "the set of all honest bankers," perhaps, or "the set of all mice weighing 20 tons." There is also a mathematical set with no members: the empty set. It is unique, because all empty sets have exactly the same number of members: none. Its symbol, introduced in 1939 by a group of mathematicians that went by the pseudonym Nicolas Bourbaki, is Ø. Set theory needs Ø for the same reason that arithmetic needs 0: things are a lot simpler if you include it. In fact, we can define the number 0 as the empty set.

What about the number 1? Intuitively, we need a set with exactly one member. Something unique. Well, the empty set is unique. So we define 1 to be the set whose only member is the empty set: in symbols, {Ø}. This is not the same as the empty set, because it has one member, whereas the empty set has none. Agreed, that member happens to be the empty set, but there is one of it. Think of a set as a paper bag containing its members. The empty set is an empty paper bag. The set whose only member is the empty set is a paper bag containing an empty paper bag. Which is different: it's got a bag in it.

The key step is to define the number 2. We need a uniquely defined set with two members. So why not use the only two sets we've mentioned so far: Ø and {Ø}? We therefore define 2 to be the set {Ø, {Ø}}. Which, thanks to our definitions, is the same as {0, 1}.

Now a pattern emerges. Define 3 as {0, 1, 2}, a set with three members, all of them already defined. Then 4 is {0, 1, 2, 3}, 5 is {0, 1, 2, 3, 4}, and so on. Everything traces back to the empty set: for instance, 3 is {Ø, {Ø}, {Ø, {Ø}}} and 4 is {Ø, {Ø}, {Ø, {Ø}}, {Ø, {Ø}, {Ø, {Ø}}}}. You don't want to see what the number of dwarfs looks like!

The building materials here are abstractions: the empty set and the act of forming a set by listing its members. But the way these sets relate to each other leads to a well-defined construction for the number system, in which each number is a specific set that intuitively has that number of members. The story doesn't stop there. Once you've defined the positive whole numbers, similar set-theoretic trickery defines negative numbers, fractions, real numbers (infinite decimals), complex numbers . . . all the way to the latest fancy mathematical concept in quantum theory or whatever.

So now you know the dreadful secret of mathematics: it's all based on nothing.

5

Voyages of discovery

Science is a journey, a process of generating ideas and testing them. And it's not always plain sailing. As John William Strutt's work demonstrates in this chapter, it can be a mystery tour needing real determination and painstaking work to stay on course. On the other hand, occasionally a single piece of intelligence can immediately illuminate a large area of terra incognita. There's an old saw which claims that "getting there is half the fun." In at least one of the following vignettes that does seem to be the case—even though it is a quest for the ultimate in boring things to do.

Absolute zero

At first blush it seems that the universe is lopsided. Light can be blindingly bright and sound deafeningly loud. But you can't get darker than dark or quieter than silence. Likewise, while stars can burn at 1 billion °C, there's a temperature below which the universe cannot go. There's a single reason for all these things—the nature of energy—yet it took centuries to understand what was really going on. Physicist Michael de Podesta picks up the story.

Absolute zero is an ideal and unattainably perfect state of coldness—the ultimate in cool. Since the concept first emerged in the mid-19th century, people have been driven to get ever closer to it. Along the way they have uncovered states of unparalleled beauty and order, developed engineering marvels and enhanced scientific insight, not least about notions of temperature and matter itself.

The idea of temperature is something we become familiar with at an early age. It is a parental rite to ensure a baby's room is stiflingly warm, its bathwater is "just right," and that it learns that some things are "Hot! Don't touch." Later on we associate numbers with different temperature sensations, and learn that 20 °C describes a warm day and 37 °C is a biochemical Mecca.

This familiarity makes it difficult to appreciate what an astonishing concept temperature embodies. Yet if you approach it with the naiveté of early natural philosophers such as Galileo, Newton and Robert Boyle, you will no longer laugh at early notions of heat. Some thought it a kind of fluid, called caloric, and we still speak of heat "flowing." Others thought that cold was caused by the presence of coldness, sometimes envisaged as "frigorific atoms." To the untutored eye, are these ideas any more absurd than the notion that light is a wave?

One experiment performed back in 1791 by Swiss physicist Marc-Auguste Pictet illustrates how baffling even simple things must have seemed. Pictet used two parabolic mirrors facing each other 21 meters apart. Each mirror reflects the light that hits it toward a focal point. He placed a thermometer at the focus of one mirror and a hot object at the focus of the other. The thermometer showed a rise in temperature indicating that "calorific rays" of some kind were being transmitted—an impressive experiment. More amazing, when snow was placed at the first focus, the thermometer reading fell several degrees. Witnesses at the time were reluctant to conclude that snow emitted cooling frigorific rays, but given knowledge at the time it would have been hard to conclude anything else!

Early efforts at measuring temperature were purely empirical. There existed a standardized method—within each laboratory at least—for determining "degrees of heat" in a reproducible manner. The most useful thermometers exploited the thermal expansion of liquids constrained in glass bulbs and narrow tubes. The level of the liquid was marked at two "fixed temperatures," such as the melting and freezing temperatures of water. Then, unknown temperatures were measured as "degrees of heat" that were etched as a scale between the two fixed points.

The biggest problem for early workers was a "thermal catch-22." The scale-marking process assumes that the liquid expands an equal amount for every unit rise in temperature. But this assumption cannot be verified unless one measures the thermal expansion of the liquid, and to do that one requires . . . a thermometer.

By the early 19th century, no solution to this circularity was in sight. Instead, different workers simply asserted that one thermometer or another was better than the others. Early thermometers used "spirit"—essentially brandy—and this was generally inferior to mercury. However, exhaustive comparisons in the 1840s by French scientist Henri Victor Regnault showed that an "air thermometer"—which measures changes in pressure of dry air in a sealed container—was superior to both in its reproducibility and inter-comparability.

Different designs of air thermometer calibrated at the freezing and boiling points of water gave consistent estimates of temperatures. In contrast, liquid-in-glass thermometers varied in their performance depending on the properties of glass, and the type of liquid. Slowly the air thermometer, which was difficult to use, began to be viewed as definitive and was used to calibrate other, more practical thermometers.

Crude as early measurements were, they brought some order to the thermal world. Reproducible readings aided everything from cooking to industrial processes. But still no one really knew what it was they were measuring!

As practical confusion lessened, theorists could turn their attention to this problem. And William Thomson, later to become Lord Kelvin, focused on the possibility of constructing a temperature scale that did not depend on the materials from which thermometers were made—an absolute scale of temperature. Kelvin's recipe for an absolute temperature scale was obscure, resting on the operation of an ideal heat engine, first imagined by the French scientist Nicolas Léonard Sadi Carnot. But a more powerful and ultimately successful "meme" was emerging: the explanation of the physical properties of matter in terms of atoms.

It is hard to imagine a time when even the greatest

scientific pioneers did not understand that everyday objects are made of atoms, that heat is the kinetic energy of moving atoms, and that temperature is a measure of the speed with which atoms move—specifically the square of the average molecular speed. Although ideas of this kind were advanced by the likes of John Herapath in 1820 and John James Waterson in 1845, they were roundly rejected by London's Royal Society. Yet by the time the book *Heat: A mode of motion* was published by John Tyndall in 1865 the idea was taught as fact.

To put this advance into a modern context, consider recent discussions about the existence of the Higgs particle. This particle is supposed to give rise to the property of matter we call "mass"—a property so familiar that most people barely think it needs explanation. Similarly the idea that the motion of hypothetical atoms was the source of heat was posited but unconfirmed for many years. The idea that heat needed a microscopic explanation was not obvious, but once established it offered astonishing insight into the role of atoms in everyday life: when we feel the temperature of a substance we are literally sensing the "buzzing" of matter.

And once the idea of molecules jiggling within a substance is accepted, the concept of absolute zero becomes inevitable: it is the temperature at which atoms become completely still.

The Fahrenheit and Celsius temperature scales assign arbitrary numbers to different phenomena. Daniel Fahrenheit used the freezing temperature of brine as a "zero" because it was the coldest temperature he could achieve. How could we hope to identify the location of absolute zero if we couldn't get close to it? Clues were around for those who knew where to look.

Guillaume Amontons, a 17th-century French instrument-maker, investigated the way the pressure of gas sealed in a vessel changed with temperature. He noted that the pressure fell by "around a quarter" when the gas was cooled from the boiling point of water to around the ice temperature. He then speculated that if cooled further, the pressure might eventually disappear. This would happen, he calculated, at what we would now describe as –300 °C, which is not far off! Later experiments of a conceptually similar kind refined this answer.

Scientists now use two temperature scales, the familiar Celsius scale and the Kelvin scale; the magnitude of a degree is the same on both scales. The Kelvin scale starts at 0 K, which translates as –273.15 °C. The melting temperature of ice (0 °C) is at 273.15 K.

With the concept of absolute temperature becoming clearer, and the possible location of "zero" identified, the race to reach zero mirrored the race to Earth's poles—a journey into the unknown.

One gas after another was cooled under pressure before being allowed to expand rapidly, which lowered its temperature further and condensed it like steam on a window. Using cascades of gases, Louis-Paul Cailletet liquefied oxygen at –183 °C and nitrogen at –196 °C. (It is doubtful whether scientists realized at this point how commonplace liquid oxygen and liquid nitrogen would become in the 20th century. Applications may have been envisaged at that time, but I would bet that making instant ice cream and destroying warts would not have been among them.)

The penultimate conquest was hydrogen by James Dewar in 1898 at –250 °C. The race to liquefy helium—the most incondensable of gases—was won by Dutchman Heike Kamerlingh Onnes at the University of Leiden, who on July 10, 1908 reached a temperature of 4.2 K. This was an astonishing technical achievement, and while it marked the end of one race, it started another that continues today—to ever lower temperatures. The few cubic centimeters of almost perfectly transparent liquid that Kamerlingh Onnes produced that day was so precious that it must have been inconceivable that it would one day be used routinely in hospitals and laboratories.

Shortly after liquefying helium, Kamerlingh Onnes discovered that at very low temperatures metals become superconductive—their electrical resistivity falls to a value indistinguishable from zero. The change is enormous—at least 15 orders of magnitude. Superconducting technology is not as commonplace as many had hoped it would become, but it is widely used in magnetic resonance body scanners where a huge magnetic field is created by an electric current in a coil of superconducting wire.

Kamerlingh Onnes did not realize that perhaps the most astonishing low temperature phenomenon of all was taking place in front of his eyes. Through small gaps in the insulated glass vessel, the precious liquid could be seen boiling. By sucking out helium vapor from the space above the liquid, the fastest helium molecules were removed and the liquid cooled even further; the vigor of the boiling increased. And then suddenly, below what we now know was 2.17 K, the bubbling stopped and the liquid became eerily still. This phenomenon was observed on the first day that helium was liquefied, but it was years before anyone understood it. A fraction of the liquid had changed to a new state known as a superfluid which has a thermal conductivity indistinguishable from infinity—so that whenever a region of the liquid became marginally hotter and began to form a bubble, the superfluid carried the heat away before the bubble could form.

As well as two protons, the nuclei of helium usually contain two neutrons (^4He). Thousands of times rarer than this was its isotope ^3He, which has only a single neutron. It was expected that the 25 percent difference in mass would change the properties of a liquid made from ^3He rather than ^4He. But the change was more profound than imagined. The lighter atoms of ^3He condensed at 3.2 K, instead of the 4.2 K of ^4He, and once liquefied it behaved completely differently, becoming more and more viscous as the temperature fell.

The difference between ^3He and ^4He exposes insight that can only be obtained at low temperatures. Who would have guessed that the presence or absence of a neutron could so transform the physical properties of a liquid made from those atoms? It is only when random thermal vibrations are reduced that the fantastic nature of atoms themselves is manifest. I would call the properties "extraordinary" but they are not—they are completely ordinary. We are just unaware of how astonishing "ordinary" matter is.

The truth is that the world in which we live is described by quantum mechanics—the laws of Newton and Lagrange that rule our familiar classical world are only approximations. The cooling of a substance exposes the quantum mechanical nature of matter. In helium, the consequences are dramatic. The electric repulsion between helium atoms is so weak that the quantum uncertainty in the position of the atoms lets them literally swap places without having to experience the inconvenience of going around each other. This ability of atoms to swap places in a structure is a characteristic property of a liquid—it's how liquids change shape so easily. And this quantum swapping of places causes both types of helium to remain liquid to the lowest temperatures investigated at normal pressures, and they are expected to remain liquid even at absolute zero.

The properties of ^3He and ^4He can be exploited in what's called a dilution refrigerator which uses the superfluidity of ^4He to allow ^3He to behave like a gas—effectively "evaporating" into a ^4He "vacuum." With this set-up, matter can be cooled to below 0.001 K, which led to further discoveries. Tungsten became superconducting at 0.012 K and ^3He itself became a superfluid at just 0.003 K. Clearly significant physical changes are still taking place within materials even at these ultra-cold temperatures.

The quest for lower temperatures in large pieces of material has stalled on the fact that the thermal conductivity and heat capacity of all materials plummet as temperature falls. This means it takes longer and longer to remove even tiny amounts of heat from a substance. Also, any experimental technique you use to study the properties of a substance will warm it up. If a butterfly happened to find itself in a refrigerator containing a cubic centimeter of copper at 0.001 K, the very act of the butterfly falling 10 centimeters would raise the copper's temperature 100-fold.

For smaller amounts of material—up to just a million atoms or so—we can cool them atom by atom using laser light. This has slowed atoms from moving at around 1 meter per second at 1 millikelvin to roughly 1 millimeter per second at 1 nanokelvin. Although applications of such technology seem unlikely at the moment, given the last century of progress we would be unwise to bet against future widespread application.

This state-of-the-art technique can get us very close to absolute zero, and I don't doubt that we will eventually get colder still. Which raises the most common question asked of cryogenic scientists: why can't we reach absolute zero? The impossibility of cooling an object to absolute zero is the essence of the Third Law of Thermodynamics, and there is no way around this.

Here's one way to understand why: conventional fridges work by placing a target to be cooled in "thermal contact" with a cooler substance, typically a recirculating fluid. We know that the fluid must be colder than the target so that heat can flow from the target. By the same principle, to get heat flowing out of a target that you want to reach absolute zero, the fluid coolant would have to be colder than 0 K to begin with! Being below absolute zero is—of course—nonsense: it is clearly impossible to make molecules move slower than not moving at all.

Techniques such as laser cooling seem to overcome the limitation of conventional cooling by simply damping the motion of atoms, but in fact all that has changed is the level of sophistication of the coolant. Even at 1 nanokelvin, atoms are moving at about 1 millimeter per second—slow, but still a long way from stationary.

It might seem odd that a century after Kamerlingh Onnes took us to 4.2 K, we are still investigating what happens in those few degrees above absolute zero. But this is perhaps because slowing down the vibration of atoms creates the equivalent of a quiet room in which one can hear tiny noises, and the logarithmic scale of the decibel—which we use to measure sound levels—could also describe the realm of cryogenic investigation. We shouldn't think about the single degree between 1 K and absolute zero, but about the factor 1,000 difference in temperature between 1 K and 1 milliKelvin. Cooling through this range, one encounters as many changes in properties as in the change from 1 K to 1,000 K.

For each factor of 10 we cool a substance, we probe atomic interactions at a new level of subtlety. So even at 1 nanokelvin, there is plenty of room for further cooling—to picokelvin, femtokelvin and beyond. And we really have no idea what we will find when we get there!

If you want to read more about the strange world around absolute zero, go to "The world of superstuff" on page 207.

Boring-ology: a happy tedium

Need something to do? Why not try watching paint dry or grass grow: at least they're better than doing nothing. You think that's a joke? Intrepid reporter Valerie Jamieson set off to discover how tedious these activities really are. In the process she discovered a whole new field of science.

Monday

Rain clouds roll ominously overhead, the wind plasters my hair across my face and I wonder what I have done to deserve this. I am slowly sinking into a muddy field just outside the Welsh seaside town of Aberystwyth. I have tied plastic bags round my feet to keep my shoes clean. I am cold, tired and, to be honest, a little bit bored. But that's the point: this is the first stop in my quest to find the most boring thing on Earth.

From the warmth and comfort of the *New Scientist* office, it all sounded like a bit of light-hearted fun. Just how tedious is watching paint dry? Does ditchwater deserve its dreary reputation? How I laughed when some smarty-pants called it boringology. Little did I know that I would be the one to draw the short straw, but here I am in a field at the Institute of Biological, Environmental and Rural Sciences (IBERS), watching grass grow.

As soon as Danny Thorogood, a turf-grass breeder here, leads me into the middle of the field I realize that not all grass is equal. Stretching in front of us are rows of different grasses that Thorogood and his colleagues have bred to be more nutritious for cows, to resist droughts, or simply to stay green. In the distance I spot giant miscanthus waving in the wind, a hybrid grass whose dry, leafless stems are a promising biofuel. Miscanthus grows at the impressive rate of 4 meters a year. "You can even hear it growing," says Mervyn Humphreys, a plant breeder at IBERS. "It crackles."

I don't know what comes over me. Suddenly I am on my hands and knees stroking the plants, examining their length, texture and color. The diversity is remarkable: the AberNile "stay green" grass is lush without the slightest hint of brown, while the bluegrass favored by North American gardeners is dark green and bushy. "For parks and lawns, you want dense coverage that doesn't grow too fast," says Thorogood. "But for grazing, you want a grass that grows quickly."

There are more than 9,000 known species of grass, but they are united in one thing, Thorogood tells me: how they grow. Unlike many other plants that sprout new shoots from the tops of mature stems, grass grows from the bottom up. Grass's growth happens near ground level in embryonic tissue called the meristem. As the plant absorbs nutrients and water, the meristematic cells divide and multiply. The cells expand as they mature, pushing older ones upward like toothpaste squeezing out of a tube. That's why mowing your lawn doesn't stop it growing—unless you scalp it to within a centimeter high and damage the meristem.

Of course, none of this means that it's interesting to watch grass grow. But I'm already suspecting that the people here find it far from dull. In fact, some of them have invented a way to measure just how fast the growth happens. "You can't just lie in a field and measure it with a ruler," says plant scientist Helen Ougham. "That would be silly."

Nearly 27 years ago, Ougham and her colleagues at what was then the Welsh Plant Breeding Station needed a controlled way to study how cooling and heating the meristem affects growth. To do this, they plucked a grass seedling from the greenhouse and sandwiched its meristem between brass plates heated or cooled with ethylene glycol, an ingredient in antifreeze. Next they clamped the youngest leaves between the jaws of a crocodile clip attached to a string looped round a pulley. To keep the string taut, they tied a counterweight to the end of it.

In the warmth of the laboratory, I get the chance to try it out for myself with a darnel grass seedling. As the plant grows, the dangling counterweight descends an equivalent distance. To measure this fall, we tie an iron cylinder halfway along the string and place it inside a "displacement transducer" that converts imperceptible movements into voltages.

Then we wait. And wait. I stifle a yawn and glance surreptitiously at my watch. Surely nothing is going to happen. The darnel grass seedling has a different idea. Within minutes, the digital voltmeter flickers into life. Grass is growing in front of my very eyes.

Every hour, it grows another 3.5 millimeters. If the temperature stays steady, my seedling will be standing over 17 millimeters taller by the time I get home tonight. Bizarrely, I am brimming with pride.

Tuesday

Yesterday I spent ten hours on a train, just for the chance to watch grass grow—and I don't regret a moment. How can ditchwater measure up to that? At first, Jane Fisher is not very confident that it can. "Ditches are not very glamorous," she admits.

I am at the Center for Ecology and Hydrology in

Wallingford, near Oxford. Fisher, a freshwater ecologist and a specialist in algae who has since moved to John Moore's University in Liverpool, has already done the dirty work for me. She has filled two jars with water taken from ditches that run into the river Thames.

I begin to sense that Fisher is warming to the boringology challenge. The air is filled with the powerful stench of manure but that, she says, is partly what makes ditches so fascinating: nutrients from the fields leach into the water, making them a rich food source for all sorts of flora and fauna. "The diversity per milliliter is huge," she enthuses.

And it turns out she's right: I can already see movement in the first jar of ditchwater. Aside from a few roots and the odd dead leaf, the water is surprisingly clear. This water comes from a ditch that runs through woodlands, and trees soak up many of the nutrients. But there is still plenty of food left over for the "ditchlife." A water snail inches up the side of the jar and a white streak zips past. It is probably a cyclops, a type of zooplankton. These strange creatures swim around grazing on green, soupy algae, removing nutrients and squirting out pellets of excrement that sink to the bottom of the ditch as sediment. Zooplankton are the reason the water is so clear.

I am really hoping to see a water bear, the toughest animal on earth. Water bears can withstand crushing pressures, shrug off lethal radiation and survive being boiled alive or chilled to near-absolute zero. They do this by completely shutting down their metabolism and then coming back to life. And I've heard that they look quite cute as they swim by, thrashing the water with their eight paws.

We place a droplet under the microscope and I peer

into an alien world. Treading water is a Keratella rotifer, a transparent microscopic creature shaped like a rectangle with spiny corners. Its mouth is covered in rotating hairs called cilia that draw water in and strain it for algae. More than 2,000 species of rotifer have been identified and they come in all shapes and sizes. But it's not just their strange appearance that makes them so fascinating to biologists.

Most species are all-female. Bdelloid or "leech-like" rotifers give birth to their young without ever having sex, so they have no need for males in their lives. How rotifers have survived so long without sex has baffled evolutionary biologists: most species that spawn offspring that are clones of themselves become extinct within a few hundred thousand years, but rotifers have been around for 70 million years.

Watching the whirr of the rotifer's cilia and seeing its internal organs is mesmerising. But my concentration is broken by blobs of algae that are bouncing around the field of view like manic ping-pong balls. These microscopic plants, Trachelomonas, are swimming around anxiously trying to move into the light to photosynthesize, Fisher tells me.

But to be honest, I'm not really listening. A nematode has just swum into view. This roundworm is less than a millimeter long and is feeding on invisible bacteria from rotting leaves. I shouldn't be surprised to see one, apparently: nematodes turn up anywhere moist, and there's even a species that loves beer mats.

There's no water bear, however, and I move on to the other jar of ditchwater. This one is much murkier and full of detritus. Under the microscope, Fisher points out the culprits, long strands of cyanobacteria and colonies of four green plant cells stuck together called Scenedesmus. They are there because the water has drained from a cow field, and is rich in nutrients. Not only do the cows' hoofs churn up the earth and release extra nutrients from the soil, but the cow dung is a rich source of phosphorus that seeps into the water.

The water is also full of diatoms, single-celled algae that convert light and nutrients into elaborate glassy shells. The ones I am staring at look like transparent coffee beans. Although they are plants, they move through the water by oozing slime from the slit in their shells. Next time you slip on a rock, you can blame it on diatoms' shiny shells and excretions.

Thursday

When planning this week's excursions, I gave myself a day off in the middle, just in case I needed to recover from the tedium. So far, however, boringology has failed to bore me.

That may be about to change. I'm heading for the Oxford office of Infinitesima, a specialist imaging company whose press releases exclaim "We really can watch paint dry!" To me, that sounds like one long bore-fest. But when I told Celia Taylor, formerly of AkzoNobel paints, what I'm going to Oxford to do, she warned me to pay attention from the start. "Most of the exciting stuff is done in twenty minutes," she says.

Exciting stuff? What she means by that is the process of forming a film. Emulsion paint, for instance, consists mainly of binder, millions upon millions of acrylic polymer particles dispersed in water. As the water evaporates, the particles merge, packing together like a stack of oranges on a fruit stall, with water filling the gaps. As the paint dries further, the particles squash together until they coalesce into a film.

Chemists like Taylor use all sorts of techniques to watch processes like these. They are always looking for ways to make paint tougher, more environmentally friendly and sport new finishes, and it all comes down to understanding paint chemistry and what happens when the water (or organic solvents, in the case of gloss paints) evaporates. Taylor's toolkit includes mass spectrometers, which sniff the molecules given off as gloss dries, and magnetic resonance imaging, which measures the amount of water left in emulsion.

But to actually see the all-important paint surface, you need something special: an atomic force microscope (AFM). This works in much the same way that a record player's needle runs along the grooves on a vinyl disc. The AFM builds up an image of individual molecules on a surface by feeling its way around with a sharp tip less than 10 nanometers wide. Making such images is a painstaking task, though. And I could miss some of the action in the minute or so it takes.

Which is why I'm here: Infinitesima's VideoAFM works at 1,000 times the speed of traditional machines, producing video images at 15 frames per second. *Paint Drying: The movie* may not be this year's Christmas blockbuster, but plenty of people pay good money to watch it.

The king of boringology

Next time you are in the bath, ponder this. Your thumbnails are growing approximately a tenth of a millimeter a day. You can thank the late American physician William Bean for that nugget of information.

Born in 1909, Bean should perhaps be crowned the founding father of boringology. His study of his own fingernails culminated in a paper published in 1980 called "Nail growth: 35 years of observation."[1]

Bean began his analysis when he was 32 by filing a horizontal line just above the cuticle of his left thumbnail. He then recorded how long it took for the mark to reach the tip of his finger. From this, he worked out that his nail grew on average 0.123 millimeters a day. Or, if you prefer, 1.4 nanometers a second.

As head of the department of internal medicine at the University of Iowa, Bean dutifully marked his thumbnail and jotted down his measurements for the next 35 years and published papers after the first 25 and 30 years. It didn't matter where Bean was—his nails grew at the same steady rate all year round.

Only two factors slowed down the growth of his talons: fungal infections and advancing years. By the age of 61, his thumbnail had slowed to 0.100 millimeters a day. And in his final paper on the subject six years later, his nails had decelerated by another 0.005 millimeters a day.

Sadly, though, when I arrive at Infinitesima, no one is watching paint dry. Instead, the VideoAFM is being used to study a molten polymer crystallising. But it gives me a flavor for what molecular-scale movies of paint drying would look like. Before my eyes, I see molecules creeping across the display. OK, so it isn't *Harry Potter and the Goblet of Fire*, and there's not much more I can say about it. But I am watching molecules move around on a surface, for crying out loud. That's pretty cool.

Maybe there's something wrong with me, but this really hasn't been the dullest week of my life. In fact, I'm rather stimulated and looking forward to regaling my friends with fascinating facts this weekend. They'll surely be mesmerized by the fact that cows fed on clover produce milk bursting with healthy polyunsaturated fats. And there's a species of parasitic nematode that can grow more than 10 meters long in sperm whales, while another species lives only in vinegar. And paint continues to harden for a whole week after it dries . . . Hang on, I'm not boring you, am I?

Putting the idle to work

Here's a conundrum. How do you find an element that does nothing—that doesn't interact with anything else? It's like trying to solve the perfect murder when there are no witnesses, almost no forensic evidence, and no body. In such cases you need two things: an alert mind and dogged determination. These ingredients led to the discovery of not one element but six. Cosmochemist David E. Fisher elaborates.

Noble gases are so called because, like the nobility, they do nothing. You might also call them rare gases, because they are so rare on Earth as to be nearly non-existent. The one exception is argon, which we inhale as 1 percent of every breath, though it has no effect on our bodies whatsoever. Helium, neon, argon, krypton, xenon and radioactive radon are odorless, tasteless, practically non-reactive wisps of unconnected atoms. In this material universe, they amount to just about nothing at all.

And yet . . . it would be hard to make a case that any other group of elements has had a greater impact on our understanding of the universe. For example, Darwin's theory of evolution needs an Earth many millions of years old in order for it to have had time to work. Yet the Bible placed a limit on Earth's age at a mere 6,000 years. How was this argument resolved? The answer was helium, which is generated in rocks containing uranium and thorium.

When these elements undergo radioactive decay they release alpha particles, which are really just helium nuclei that easily pick up electrons to create the gas. In 1906, armed with this idea and the rate of production of alpha particles by uranium, thorium and their decay products, Ernest Rutherford and Frederick Soddy dated several rocks at up to 500 million years; Earth would have to be at least that old. (Later work with lead isotopes pinned down the age to around 4.5 billion years.) Not only did Rutherford and Soddy create the concept of radioactive dating, they also kick-started our modern understanding of the cosmos and its great age.

What if you want to probe the interior of the sun? The answer is to use argon, as physicist Ray Davis did. He focused on solar neutrinos—ghostly particles created by nuclear fusion in the sun's core—as a way to test models of nuclear reactions in stars. Neutrinos reverse the natural decay of the radioactive isotope argon-37 into chlorine-37. So in 1958, Davis set up a huge vat of cleaning fluid containing chlorine-37 deep in a mine in South Dakota and used a Geiger counter to detect any argon created. Davis's pioneering work revealed much about not only the sun but also the peculiar nature of neutrinos. It won him the Nobel prize for physics 44 years later.

Xenon, meanwhile, can tell you about the formation of the solar system. Xenon-129 is an isotope produced by the radioactive decay of iodine-129, which is created in quantity only in supernovas and has the relatively short half-life, in cosmological terms, of 16 million years. The discovery of unexpectedly large amounts of xenon-129 in meteorites was the first evidence that the solid bodies of the solar system formed within the surprisingly short time of a hundred million years after a nearby supernova seeded the material that made them. That shocked theorists who thought it could never have happened so quickly.

Even though the noble gases are rare on Earth, they are not rare in the universe as a whole. This tells us that Earth's atmosphere must have formed after the planet itself. As Earth formed it was too small to retain gases, which drifted away into the cosmos. The main components of today's atmosphere—nitrogen, oxygen, water and carbon dioxide—must have been locked away in non-volatile forms. Water was trapped in hydrated minerals, carbon dioxide in carbonates, and so on. Only as the Earth and its gravitational attraction grew did these gases, escaping from volcanic eruptions, create the atmosphere.

Finding roles for krypton and friends

They may be lazy loners, but the noble gases have found useful roles. Just think how dull it would be downtown without the red glow of neon lights or the blue-white of krypton. They play more profound roles, too. Superconductivity, for example, was discovered while searching for the coldest temperatures on Earth using liquid helium.

In the Second World War, the Allies wanted to know how Hitler's attempts to build an atomic bomb were going. So they attached a trap beneath a bomber and flew it over suspect German sites in search of xenon-133. This is a fission product of uranium that doesn't react with anything else and has a half-life of five days, so should hang around long enough to be detected. A positive result would have been definitive, but the negative result they obtained meant that they were looking at the wrong sites, or the experiment was somehow flawed, or—as proved to be the case—Hitler didn't have the bomb.

Xenon-133 is also valuable in medicine. It is used as a radioactive marker to identify pulmonary embolisms, and xenon gas is an excellent anesthetic, and is used today in Russia and Germany.

So these wisps of nearly nothing reveal much about Earth and its place in the universe. Yet for me the most fascinating aspect of the noble gases is how they were discovered. By the 1860s more than fifty elements had been found, often revealing themselves when subjected to the actions of other chemicals, heat or even electricity. We now know that the noble gases are, in the main, stubbornly non-reactive because they contain a full outer shell of electrons—a prerequisite for stability. But back in mid-Victorian times their aloofness meant the noble gases had completely eluded detection.

The first hint of their existence appeared in 1868 as a faint line in the spectrum of light from the sun, indicating the presence of an element not known on Earth. This was given the name helium, after Helios, the Greek god of the sun. At the time, it raised speculation about elements in the stars being different from those on Earth, but a few years later the same line was found when a uranium mineral called cleveite was heated, and the Earth and sun were once more united.

Nothing happened for a while, until the trail was picked up from a different direction with a different end in view, when the British physicist John William Strutt—himself a noble, Lord Rayleigh—began to wonder why the atomic weights of the elements seem to be nearly whole-number multiples of hydrogen. Why whole numbers? And even more puzzling, why only "nearly" whole numbers?

His attitude was, if you don't understand something, measure it. He spent ten years making precise measurements of the densities of the gases, from which their atomic weights could be calculated, starting with hydrogen, oxygen and then nitrogen. No reason to expect a breakthrough here; it was a routine experiment. Rayleigh bubbled air through liquid ammonia, NH_3, and then passed it through a tube containing red-hot copper. That stripped the air of its oxygen, which combined with hydrogen from the ammonia, leaving just nitrogen.

Rayleigh did what a good scientist does: he carried out this experiment again and again to check the results. He then repeated it with a difference. Initially, some of his nitrogen would have come from the ammonia he used; this time he got rid of the ammonia so all the nitrogen came from air. "To my surprise and disgust the densities of the two methods differed by a thousandth part," he wrote.

Nitrogen from air was apparently heavier than that from ammonia by just 0.1 percent. I would have put it down to experimental error and moved on. But as Rayleigh said, "It is a good rule in experimental work to seek to magnify a discrepancy when it first appears rather than to follow the natural instinct to trying [*sic*] to get quit of it."

That's just what he did, this time replacing air with oxygen so that the nitrogen he collected came only from ammonia. He found that the discrepancy was indeed magnified: it was now 0.5 percent. Something real was happening, but what? He wrote a letter to the journal *Nature*, asking for help. It began, "I am much puzzled by some recent results as to the density of nitrogen, and shall be obliged if any of your chemical readers can offer suggestions as to the cause."

First suggestion: nitrogen in air is nothing but nitrogen, while nitrogen in ammonia is chemically combined with hydrogen. So perhaps he had nitrogen in two different chemical states which affected their atomic weights. But how?

No answer. Bad idea. Start again.

Finally, after other suggestions led nowhere, came the idea that a heavier gas might be mixed in with nitrogen from air. This contradicted Occam's razor, which in colloquial terms means "keep it simple." Invoking an unknown substance, a cryptic gas heavier than nitrogen, to explain the results had shades of phlogiston and the ether—illusory substances invented in other contexts as a fig-leaf for our lack of understanding.

But there is an even more hallowed tenet of science: test your ideas, experiment and observe. So in 1894, together with William Ramsay, Rayleigh passed electrical sparks through air augmented with pure oxygen to produce nitrogen oxides. They removed these by dissolving them in a weak alkali solution. Lo and behold, when all the nitrogen and oxygen were gone, a small amount of colorless gas remained, which they named argon. That comes from the ancient Greek for a lazy thing, since the gas wouldn't react with anything. It showed a pattern of emission lines never seen before, so argon was not only a previously unsuspected component of air, it was an entirely new element.

Ramsay moved on to investigate whether the gas seeping out of uranium-bearing rocks was argon, but in 1895 identified it as helium. Arguing from his understanding of the then-primitive periodic table, he suggested that helium and argon might represent a new family of elements. He went so far as to predict another such element with a mass of 20. He soon discovered it and named it neon. Krypton and xenon followed several years later, and in 1904 both men received a Nobel prize, Rayleigh in physics and Ramsay in chemistry. This is the only time an element or column of elements has been the basis for these two prizes in the same year.

In 1910, Ramsay collected the full set by producing and characterising radon. This nasty radioactive gas had been noticed before, but it was Ramsay who proposed and then demonstrated that it was another noble gas.

The discovery of the noble gases fascinates me because it is about the whole fabric of science and the roots of discovery. Rayleigh was not looking for a new element, he was trying to solve the riddle of nearly whole-number atomic weights. In this he failed: the explanation awaited the discovery of both protons and neutrons. The discovery of argon, which opened the door to the other noble gases, was serendipitous, the result of chance combined with careful experimentation and an open, inquiring mind. Like many other important scientific advances, it happened not as a result of purposeful planning, but while trying to understand something else.

So, if you want to succeed in science, keep in mind the advice offered by cosmochemist Michael Lipschutz to his students: "Obey the Biblical injunction: seek and ye shall find. But seek not to find that for which ye seek."

Impossible reaction

If there is one half-remembered chemical fact that most of us carry from our schooldays, it is that the inert or noble gases do not react.

The early history of these elements, which are ranged in the right-hand column of the periodic table, provided ample support for that view. Just after the noble gas argon was discovered in 1894, the French chemist Henri Moissan mixed it with fluorine, the viciously reactive element that he had isolated in 1886, and sent sparks through the mixture for good measure. Result: nothing. In 1924, the Austrian Friedrich Paneth pronounced the consensus. "The unreactivity of the noble gas elements belongs to the surest of all experimental results," he wrote. The theory of chemical bonding explained why. The noble gases have full outer shells of electrons, and so cannot share other atoms' electrons to form bonds.

The influential chemist Linus Pauling was one of the chief architects of that theory, yet he didn't give up on the noble gases immediately. In the 1930s, he managed to get hold of a rare sample of xenon and persuaded his colleague Don Yost at the California Institute of Technology in Pasadena to try to get it to react with fluorine. After much cooking and sparking, Yost succeeded only in corroding the walls of his supposedly inert quartz flasks.

After that, it was a brave or foolish soul who still tried to make noble-gas compounds. The late British chemist Neil Bartlett, working at the University of British Columbia in Vancouver, was not trying to defy conventional wisdom, he was just following common logic.

In 1961, he discovered that the compound platinum hexafluoride (PtF_6), first made three years earlier by US chemists, was an eye-wateringly powerful oxidant. Oxidation, the process of removing electrons from a chemical element or compound, bears oxygen's name because oxygen has an almost unparalleled ability to perform the deed. But Bartlett found that PtF_6 could even oxidize oxygen, ripping away its electrons to create a positively charged ion.

Early the next year, Bartlett was preparing a lecture and happened to glance at a textbook graph of "ionization potentials." These numbers quantify the amount of energy required to remove an electron from various substances. He noticed that xenon's ionization potential was almost exactly the same as oxygen's. If PtF_6 could oxidize oxygen, might it oxidize xenon, too?

Mixing red gaseous PtF_6 and colorless xenon supplied the

answer. The glass vessel was immediately covered with a yellow material. Bartlett found it to have the formula $XePtF_6$–xenon hexafluoroplatinate, the first noble-gas compound.

Other compounds of xenon and then krypton followed. Some are explosively unstable: Bartlett nearly lost an eye studying xenon dioxide. Radon, a heavier, radioactive noble gas, forms compounds too, but it wasn't until 2000 that the first argon compound, argon fluorohydride, was reported to exist at low temperatures by a group at the University of Helsinki.[1] Even now, the noble gases continue to produce surprises. Nobel laureate Roald Hoffmann of Cornell University in Ithaca, New York, admits to being shocked when, also in 2000, chemists in Berlin reported a compound of xenon and gold–the metal gold is supposed to be noble and unreactive too.

So don't believe everything you were told at school. Noble gases are still the least reactive elements out there; but it seems you can coax elements to do almost anything.

Philip Ball

To learn more about "things" that do nothing, just carry on reading!

Get up, get out of bed

In 1966, Gregg Hill took the world's laziest summer job. First he was poked and prodded and had his fitness assessed by every technique then known to medicine. Then, for 20 days, he and four other student volunteers became the ultimate couch potatoes, confined to bed—not even allowed to walk to the toilet. The goal was to investigate how astronauts would respond to space flight, but when Hill and his fellows finally staggered to their feet, their drastic deterioration helped spark a revolution in medical care here on Earth. As Rick A. Lovett explains, before the experiment took place, bed rest was recommended for people with weak hearts. Afterward, doctors knew that it made them worse.

The five men were the image of American youth, circa 1966: well groomed and confident. America was racing for the moon, but these young men were looking beyond, doing their bit for astronauts in orbiting space stations and perhaps eventually on a trip to Mars.

They had volunteered for what is now known as the Dallas Bed Rest and Training Study. The goals were twofold: to simulate the effects of weightlessness on astronauts and to determine how quickly the body recovered when normal life resumed. As an aside, the scientists who were monitoring the effects of such slothfulness hoped to find out why hospital patients feel as weak as kittens after lengthy stays in bed. Speculation at the time focused on extended inactivity causing blood to pool in the limbs, producing a dizzying drop in blood pressure when you stood up. But maybe it was something more insidious, such as changes in the heart or lungs. In 1966, nobody knew.

One of the volunteers was Gregg Hill, a college student with an interest in exercise physiology. He was also a runner who could do the mile in 4 minutes 45 seconds—admittedly not Olympic standard, but no slouch either.

Initially, the study leader, Carleton Chapman of the University of Texas Southwestern Medical School, had signed up six volunteers: three athletes and three less active students, to see how they compared. But one of the athletes, "a big handsome hunk of a swimmer," backed out, Hill says, when he discovered how many needles would be stuck in his body.

The tests were in part inspired by the findings of Archibald Hill, the British pioneer of biophysics. Forty years earlier he had discovered that during exercise the body reaches a state of maximal oxygen uptake which cannot be exceeded no matter how hard you work. If you try to go faster, you fall into what athletes call "oxygen debt," in which you can briefly sprint, but must then stop to recover.

Maximal oxygen uptake is referred to as VO_2max. The standard test monitors your oxygen consumption as you run on a treadmill, with a technician gradually turning up the grade until you have to pack it in. For competitively inclined people who try to "beat" the test, it's a brief but brutal workout.

In addition to VO_2max, Chapman's team wanted to know everything possible about the students' hearts, lungs and overall fitness. They took chest X-rays to determine the volume of their hearts. They measured lung capacity by making them exhale into a device called a spirometer. They weighed them underwater to calculate how much body fat they were carrying.

The tests that frightened off the swimmer were designed to measure the amount of blood pumped by each beat of the heart (the "stroke volume") and the fraction of oxygen removed from it by the leg muscles. Today, there are non-invasive ways to measure these, but in

1966, one needle had to be stuck into a vein in the right arm and another into an artery in the left one. Squirts of green dye were injected into the vein, and the amount by which the dye was diluted when it appeared in the artery revealed the volume of blood with which it had mixed in the heart. Measuring how much oxygen the legs were using required yet another needle to extract samples from a leg vein—all while running on a treadmill. Definitely not a test for the squeamish.

Preliminaries completed, Hill and his friends went to bed. Their diets were monitored so that they wouldn't gain weight, but exercise was strictly forbidden. The only concession was a single brief shower halfway through the experiment.

Bored, Hill and his wardmates read a lot, watched TV and listened to music—although that sometimes caused energy-consuming arguments. "I like classical," Hill says, "but I tolerated pop for the sake of peace."

When they were finally released, the men were placed on wheelchairs and wheeled to the sports lab for a repeat of the initial tests. The results were stunning. Chapman's team found that a mere three weeks of inactivity had cut VO_2max by 28 percent and stroke volume by 25 percent—more than 1 percent per day. Overall, their hearts had shrunk 11 percent, and two of the non-athletes fainted during their first efforts on the treadmill.

As word filtered out, hospital doctors began prodding surgical patients out of bed as soon as possible and cardiologists began prescribing exercise rather than bed rest for heart patients. Hill's boring summer job had changed the face of medicine.

Back on his feet, however, Hill now had to work harder, as the study entered its "training" phase. For the next 55 days, he endured intense workouts, including time trials on the track. "That was rough," he says. Early on, he even had trouble driving because his legs were so sore from the training that they trembled when he pushed the pedals. By the end, though, he and the other athletes had fully recovered, and the three non-athletes were in better shape than at the start of the study.

An academic paper, published in 1968, reports these findings in dozens of pages of charts and dry language. Hill puts it more succinctly. "The heart is a tremendously flexible organ," he says. "It remodels itself to meet changing conditions very quickly—much more quickly than muscles respond to weightlifting."

In later years, Hill maintained his interest in exercise physiology but went on to become a college instructor in computer science. Then, in 1996, he received a phone call. Was he willing to be part of a follow-up study? This time there would be no needles and no need for bed rest.

The follow-up was the brainchild of Darren McGuire and Benjamin Levine from the University of Texas Southwestern Medical Center in Dallas. Their interest wasn't in bed rest, but in the effect of age on cardiovascular fitness.

Few had looked at this before and those who had tended to focus on professional athletes, making it difficult to separate the effects of aging from those of retirement from sport. Hill's group provided a unique opportunity because no such group of relatively ordinary people had ever been so comprehensively studied for so long.

A few months later, all five men were back on the

treadmill—this time minus the needles. Today, sophisticated imaging techniques show how well the heart is functioning. Then they were put on tough training routines.

The follow-up findings, published in 2001, were nearly as spectacular as the original ones. First, McGuire and Levine discovered that 30 years of aging had taken less of a toll on Hill and the others than 20 days of bed rest. Even though all five men had lost condition (and gained weight), the decades had reduced their VO_2max only half as much as their stints in bed.

That was interesting, but more important was what happened when the men were put on exercise programmes building up to about three to five hours per week. Within six months, their VO_2max levels rebounded all the way to what they had been at the end of the 1966 study. "We reversed thirty years of aging with six months of training," Levine said.

This did not, however, give Hill back his ability to run a 4:45 mile, most probably, he suspects, because his aging tendons have lost elasticity. Still, he likes the fact that doctors and nurses often tell him he has the vital signs of a teenager. "There is a fountain of youth," he says. "It's just that you have to work hard to drink from it."

If you're interested in more things that do nothing, try "The workout pill" on page 198.

6

Conclusions

Conclusions in science are strange in that they are not for ever. "Nothing is static, nothing is final, everything is held provisionally," said Jocelyn Bell Burnell, the astronomer who discovered pulsars. Put plainly, no sooner do you get used to one theory than somebody comes up with a better one. It is often forgotten, however, that theories can be useful even when they are not correct: the prevailing theory prevails precisely because it describes reality better than any other. So here are three articles on our latest conclusions about conclusions—the stark consequence of slothful living, the raw nature of reality when it is robbed of energy, and the demise of our collective home, the universe.

The workout pill

It's an old idea that doing nothing is against the natural order. The message normally comes down from a parent or teacher: you may have been warned that idleness is incompatible with purity of mind or that the devil will commandeer your hands.

But in the last century, science showed unambiguously that inactivity is also bad for your body—today we're finding out just how bad. Andy Coghlan discovers the extraordinary value of a little daily exercise.

It's 9:00 AM in the office—time for my daily medication. As usual, I slink off to the fire escape for my fix. Twenty minutes later, I'm back at my desk, brimming with vitality and raring to go.

I've taken this medicine regularly now for years, after developing elevated blood pressure in my mid-40s. I'd heard it could help reduce blood pressure and improve circulation. Sure enough, the high blood pressure vanished long ago.

Amazingly, this drug is freely available to everyone on the planet. It's completely up to you when you take it, and how much. And as research is now revealing, the more of it you take, the healthier you will be.

What is this wonder drug? It is plain old physical activity of all sorts—from running marathons to simply walking around your sofa while watching television. We've all heard that exercise is good for us, but what is becoming increasingly clear is the sheer extent of its benefits and why it works.

A plethora of recent studies show that exercise protects us from heart attacks, strokes, diabetes, obesity, cancer, Alzheimer's disease and depression. It even boosts memory. And it has the potential to prevent more premature deaths than any other single treatment, with none of the side effects of actual medication. "It's a wonder drug," says Erik Richter, a diabetes researcher at the University of Copenhagen. "There's probably not a single organ in the body that's unaffected by it."

Throughout evolution, humans have been active. Our ancestors chased prey as hunter-gatherers and fled from predators. More recently, they laboured on farms and in factories. But the decline of agricultural and industrial labor, plus the invention of the car, a multitude of labor-saving devices and—most perniciously—TV, computers and video games, mean we've all ground to a sudden and catastrophic standstill.

"We were built to be active, but the way our environment has changed and the way we live our lives has led us to become inactive," says Christopher Hughes, senior lecturer in sport and exercise medicine at Queen Mary, University of London.

Now we're paying the price. In 2009, Steven Blair, an exercise researcher at the University of South Carolina in Columbia, published a study of more than 50,000 men and women showing that a lack of cardiorespiratory fitness was the most important risk factor for early death.[1] It accounted for about 16 percent of all deaths in men and women over the period of study, more than the combined contributions of obesity, diabetes and high cholesterol, and double the contribution of smoking.

In other words, physical inactivity is killing us. "Everyone knows too much booze or tobacco is bad for you, but if physical inactivity was packaged and sold as a product, it would need to carry a health warning label," says Hughes.

As we have become inactive, so once-rare diseases have mushroomed. A report from the organization Diabetes UK reveals that in 1935, when the world's population was just over 2 billion, an estimated 15 million people globally had type 2 diabetes.[2] By 2010 the world's population had more than tripled and the number with diabetes had shot up to 220 million—more than 14 times the number in 1935. Likewise, results published in 2012 in the *Journal of the American Medical Association* show that more than a third of men and women in the US are obese, as are about 17 percent of US children.[3]

The good news is that we can do something about it. I started running up and down the fire escape for a few minutes each day in the hope of not having to take cholesterol-lowering statins or drugs for high blood pressure. Now I'm eager to know what my daily routine is doing to my body and, more importantly, how it may be protecting me from disease.

The most robust evidence so far comes from the Exercise is Medicine initiative pioneered by the American College of Sports Medicine in Indianapolis, Indiana. Researchers there have collated studies over the past decade or so of people who follow the US government's advice on physical activity. This prescribes 150 minutes per week of moderate-intensity aerobic activity, such as brisk walking, ballroom dancing or gardening, or 75 minutes of more vigorous activity such as cycling, running or swimming.

What the Exercise is Medicine findings show is that this weekly dose of moderate exercise reduces the risk of premature death through heart disease by 40 percent, approximately the same as taking statins.

Chi Pang Wen of the National Health Research Institute in Zhunan, Taiwan, offers some insights into precisely how physical activity prevents cardiovascular diseases.

"Exercise can stimulate circulation, flush out fatty deposits in the walls of blood vessels and dilate small vessels that could otherwise be the cause of a heart attack or stroke," he says. In April 2012 he presented results from a study of more than 430,000 Taiwanese men and women, showing that exercise reduced the risk of heart attacks by 30 to 50 percent.

Exercise also keeps blood vessels clear by helping to destroy the most dangerous fats. Research published earlier in 2012 reveals that it alters the structure of fatty triglyceride particles in the bloodstream, making it easier for enzymes to destroy them before they can gum up the works.[4] Many risks to circulatory health come from such fatty particles, in the form of chylomicrons produced in the gut, or very low density lipoproteins (VLDLs) pumped out by the liver. The bigger the VLDL particles are, the easier they are for enzymes to break down, and the findings show that exercise causes the particles to enlarge by about a quarter.

"A single 2-hour bout of exercise reduced triglyceride concentrations in the circulation by 25 percent compared with no exercise," says Jason Gill, who led the study at the University of Glasgow. His team found a decrease in both types of fat, but the decrease was twice as much for the more insidious VLDL particles.

One of the most startling findings of the Exercise is Medicine initiative is that a modest weekly dose of exercise lowers the chances of developing type 2 diabetes by 58 percent, twice the preventive power of the most widely prescribed anti-diabetes medication, metformin.[5]

Type 2 diabetes affects adults when they stop

responding efficiently to the hormone insulin, which orders muscle and fat cells to absorb surplus glucose from the bloodstream. When insulin loses its punch, glucose continues circulating and creates the potentially fatal sugar imbalances that are the hallmark of diabetes.

How does exercise reverse this? The story dates back to 1982, when Richter found that insulin activity is enhanced by physical activity—at least in rats.[6] Experiments showed that after the rats had run around for a couple of hours, their cells became up to 50 percent more responsive to insulin compared with the cells of non-exercising rats. "We confirmed it later in humans," Richter says.

As cells reawaken to insulin, it seems that surplus glucose gets sponged from the circulation. Richter found that the effects lasted for a couple of hours after exercise in rats, and up to two days in humans.[7]

Recently he and colleagues have unraveled more details about how exercise brings this about. They have discovered that both insulin and muscle contractions during exercise activate a molecule in muscle and fat cells called AS160, which helps them absorb glucose.[8] Once activated, AS160 orders the cell to send molecules to the cell's surface to collect glucose and bring it inside. Without these transporter molecules, glucose cannot get through the fatty cell membrane.

Exercise also helps cells burn off excess sugar. Muscle cells absorb glucose and fatty acids from the bloodstream to replenish adenosine triphosphate (ATP), the molecular fuel found in most living cells. As ATP is used up, it produces waste products that are sensed by another molecule, AMPK. AMPK then orders cells to recharge by absorbing and burning yet more fat and sugar. In the mid-1990s, Grahame Hardie at the University of Dundee found that exercise accelerates this process because muscle contraction activates AMPK.

Hardie says exercise has the potential to reverse obesity and diabetes and prevent cancer. The findings of the Exercise is Medicine initiative show that taking the US government's recommended weekly dose of exercise halves the risk of breast cancer in women and lowers the risk of colorectal cancer by around 60 percent.[9, 10] This is about the same reduction seen with low daily doses of aspirin.

How exercise does this is not yet clear—not least because so many factors are involved in cancer's appearance and progression, including sex hormone imbalances, the ability of the immune system to clear cancer cells, and damage to genes and DNA generally. However, some clues are beginning to emerge. "Exercise reduces body weight, which is a known risk factor for postmenopausal breast cancer," says Lauren McCullough of the University of North Carolina at Chapel Hill.

She also thinks that reducing fat deposits in the body results in less exposure to circulating hormones, growth factors and inflammatory substances. "All have been shown to raise breast cancer risk," she says.

Another clue comes from work by Anne McTiernan of the Fred Hutchinson Cancer Research Center in Seattle, who studies colorectal cancer. Biopsies from 200 healthy volunteers showed that, compared with exercisers, non-exercisers had more telltale signs of abnormalities in colonic crypts—recesses in the lining of the colon that absorb water and nutrients.[11] Crypts in idle participants had an increased number of dividing cells, and these also climbed higher up the crypt walls, where they had the potential to form pre-cancerous polyps.

Another potential protection against cancer might come back to the ability of exercise to stimulate AMPK. Recent research by Beth Levine of the University of Texas Southwestern Medical Center in Dallas showed that exercise stimulates cells craving extra energy to burn unwanted rubbish, including faulty or mutated DNA that could trigger cancer if it hangs around.[12] More recently, Levine has discovered the same processes in brain cells, suggesting that exercise might play a role in staving off dementias and neurodegeneration.

As well as potentially staving off dementia, pounding the stairs might even help boost my brainpower and memory. Back in 1999, Henriette van Praag of the US National Institute on Aging in Baltimore, Maryland, found that mice using a running wheel developed new neurons in the hippocampus, a part of the brain vital for memory.[13] "We had a doubling or tripling of neurons after they'd been running daily for about a month," she says. Subsequently, van Praag and other groups found the most likely reason: a doubling in the level of a substance in the hippocampus called brain-derived neurotrophic factor, or BDNF, which may support growth of new neurons.

More than a decade on, a team led by Art Kramer of the University of Illinois at Urbana-Champaign demonstrated through a brain-imaging study of 120 older adults that exercise increased hippocampus volume by around 2 percent.[14] It also improved their memory, as measured by standard tests. "The volume increase we saw can make up for approximately two years of normal age-related decrease," says Kramer. "We found that even modest increases in fitness can lead to moderate, 15 to 20 percent improvements in memory."

The benefits aren't just restricted to adults. Kramer and his colleagues have also found that pre-adolescent children who exercise develop larger hippocampuses.[15]

So if exercise is so beneficial, why won't people take it? At least 56 percent of US adults don't meet the government's exercise guidelines. "The most common excuse people give in polls is that they don't have time," says Blair. Perhaps that is not surprising when US citizens spend, on average, almost 8 hours a day watching TV, according to a 2008 study.

For those, like me, who don't want the fuss of joining a gym, there is plenty people can do at home or the workplace in their own time and at their own pace. Blair cites a study in which researchers asked half of a group of couch potatoes to walk around their sofa during each TV commercial break.[16] "They burned 65 calories more per hour, and that is 260 calories in 4 hours," he says. Over a week, their exertions met the US government recommendations for exercise.

And overweight people can benefit massively from exercise even if they don't lose weight, Blair points out. One of his studies has shown that for fit fat people, the risk of dying prematurely is half that for unfit lean people.[17]

Once a marathon runner, Blair now walks for an hour a day, and at the age of 73, he has set himself the goal of walking 5 million steps each year, tracking his progress with a pedometer. He is concerned that not enough doctors recognize that lack of fitness is effectively a disease. He wants them to use fitness as a gauge of health, perhaps making their patients do a treadmill test as a matter of routine, rather than considering it as an afterthought.

Figures published in *The Lancet* in 2012 back up his assertion that no action, other than abstaining from smoking, is as good for health as being physically active.[18] The study also reveals that physical inactivity effectively kills 5 million people a year worldwide, as many as smoking.

As for me, the stair-run does seem to be working, although I don't have health data from eight years ago to confirm my progress. More recent scans and tests showed my blood pressure and bone density are normal, and I have 6 percent less body fat than is average for my age. Also, only 20 percent of my fat is the dangerous sort around organs in the abdomen, compared with 30 percent in most of my peers. My heart fitness, measured on a treadmill, is above average and I have no chronic diseases that I know of. Now, imagine you were offered a pill that did all that. Wouldn't you take it?

The world of superstuff

Cool matter to absolute zero and you rob it of all thermal energy. As it approaches its ultimate state, a new world opens up. It's a world in which the everyday rules of physics seem to disappear. As Michael Brooks finds out, here you glimpse nature in the raw.

For centuries, con artists have convinced the masses that it is possible to defy gravity or walk through walls. Victorian audiences gasped at tricks of levitation involving crinolined ladies hovering over tables. Even before then, fraudsters and deluded inventors were proudly displaying perpetual-motion machines that could do impossible things, such as make liquids flow uphill without consuming energy. Today, magicians still make solid rings pass through each other and become interlinked—or so it appears. But these are all cheap tricks compared with what the real world has to offer.

Cool a piece of metal or a bucket of helium to near absolute zero and, in the right conditions, you will see the metal levitating above a magnet, liquid helium flowing up the walls of its container or solids passing through each other. "We love to observe these phenomena in the lab," says Ed Hinds of Imperial College, London.

This weirdness is not mere entertainment, though. From these strange phenomena we can tease out all of chemistry and biology, find deliverance from our energy crisis and perhaps even unveil the ultimate nature of the universe. Welcome to the world of superstuff.

This world is a cold one. It only exists within a few degrees of absolute zero, the lowest temperature possible. Though you might think very little would happen in such a frozen place, nothing could be further from the truth. This is a wild, almost surreal world, worthy of Lewis Carroll.

One way to cross its threshold is to cool liquid helium to just above 2 K. The first thing you might notice is that you can set the helium rotating, and it will just keep on spinning. That's because it is now a "superfluid," a liquid state with no viscosity.

Another interesting property of a superfluid is that it will flow up the walls of its container. Lift a bucketful of superfluid helium out of a vat of the stuff, and it will flow up the sides of the bucket, over the lip and down the outside, rejoining the fluid it was taken from.

Though fascinating to watch, such gravity-defying antics are perhaps not terribly useful. Of far more practical value are the strange thermal properties of superfluid helium.

Take a normal liquid out of the refrigerator and you find it warms up. With a superfluid, though, the usual rules no longer apply. Heat diffuses so quickly through a superfluid that it doesn't warm. Researchers working at the Large Hadron Collider (LHC) at CERN, near Geneva in Switzerland, use this property to help accelerate beams of protons. They pipe 120 tons of superfluid helium around the accelerator's 27-kilometer circumference to cool the thousands of magnets that guide the particle beams. Normal liquid helium would warm up considerably if used in this way, but the extraordinary thermal properties of the superfluid version means its temperature rises by less than 0.1 K for every kilometer of the beam ring. Without superfluids, it would have been impossible to build the machine that many physicists hope will reveal the innermost secrets of the universe's forces and building blocks

The LHC magnets have super-properties themselves. They are made from the superfluid's solid cousin, the superconductor.

At temperatures approaching 0 K, many metals lose all resistance to electricity. This is not just a gradual reduction in resistance, but a dramatic drop at a specific temperature. It happens at a different temperature for each metal, and it unleashes a powerful phenomenon.

For a start, very little power is needed to make superconductors carry huge currents, which means they can generate intense magnetic fields—hence their presence at the LHC. And just as a superfluid set rotating will keep rotating for ever, so an electric current in a superconducting circuit will never fade away. That makes superconductors ideal for transporting energy, or storing it.

The cables used to transmit electricity from generators to homes lose around 10 percent of the energy they carry as heat, due to their electrical resistance. Superconducting cables would lose none.

Storing energy in a superconductor could be an even more attractive prospect. Renewable energy sources such as solar, wind or wave power generate energy at unpredictable times. If superconductors could be used to store the excess power these sources happen to produce when demand is low, the world's energy problems would be vastly reduced.

We are already putting superconductors to work. In China and Japan, experimental trains use another feature of the superconducting world: the Meissner effect.

Release a piece of superconductor above a magnet and it will hover above it rather than fall. That's because the magnet's field induces currents in the superconductor that create their own, opposing magnetic field. The mutual repulsion keeps the superconductor in the air. Put a train atop a superconductor and you have the basis of a levitating, friction-free transport system. Such "maglev" trains do not use metal superconductors, because it is too expensive to keep metals cooled to a few K; instead they use ceramics that can superconduct at much higher temperatures, which makes them easier and cheaper to cool using liquid nitrogen.

These are strange behaviors indeed, so what explains them? Superfluidity and superconductivity are products of the quantum world, and to get an idea of what's going on, here's a thought experiment. Imagine you have two identical particles, and you swap their positions. The new set-up looks exactly the same, and responds to an experiment exactly as before. However, quantum theory records the swap by a change in the particles' quantum state, which is multiplied by a "phase factor." Switching the particles again brings in the phase factor a second time, but the particles are in their original positions and so everything returns to its original state. "Since switching the particles twice brings you back to where you were, multiplying by this phase twice must do nothing at all," says John Baez at the Center for Quantum Technologies in Singapore. This means that squaring the phase must give 1, which in turn means that the phase itself can be equal to 1 or −1.

This is more than a mathematical trick: it leads nature to divide into two. According to quantum mechanics, a particle can exist in many places at once and move in more than one direction at a time. In the last century, theorists showed that the physical properties of a quantum object depend on summing together all these possibilities to give the probability of finding the object in a certain state.

There are two outcomes of such a sum, one where the phase factor is 1 and one where it is –1. These numbers represent two types of particles, known as bosons and fermions respectively.

The difference between them becomes clear at low temperatures. That is because when you take away all thermal energy, as you do near absolute zero, there aren't many different energy states available. The only change that can be made is to swap the positions of the particles, with the consequent phase change.

Swapping bosons introduces a phase change of 1. Using the equations to work out the physical properties of bosons, you find that their states add together in a straightforward way, and that this means there is a high probability of finding indistinguishable bosons in the same quantum state. Simply put, bosons like to socialize.

In 1924, Albert Einstein and Satyendra Bose suggested that at low enough temperatures, the body of indistinguishable bosons would effectively coalesce together into what looks and behaves like a single object, now known as a Bose–Einstein condensate, or BEC.

Helium atoms are bosons, and their formation into a BEC is what gives rise to superfluidity. You can think of the helium BEC as a giant atom in its lowest possible quantum energy state. Its strange properties derive from this.

The lack of viscosity, for instance, comes from the fact that there is a huge gap in energy between this lowest state and the next energy state. Viscosity is just the dissipation of energy due to friction, but since the BEC is in its lowest state already, there is no way for it to lose energy—and thus it has no viscosity. Only by adding lots of energy can you break a liquid out of the superfluid state.

If you physically lift a portion of the superatom, it acquires more gravitational potential energy than the rest. This is not a sustainable equilibrium for the superfluid. Instead, the superfluid will flow up and out of its container to pull itself all back to one place.

Superconductors are also BECs. Here, though, there is a complication because electrons, the particles responsible for electrical conduction, are fermions.

Fermions are loners. Swap them around and, as with swapping your left and right hand, things don't quite look the same. Mathematically, this action introduces a phase change of –1 into the equation that describes their properties. The upshot is that when it comes to summing up all the states, you get zero. There is zero probability of finding them in the same quantum state.

We should be glad of this: it is the reason for our existence. The whole of chemistry stems from this principle that identical fermions cannot be in the same quantum state. It forces an atom's electrons to occupy positions further and further away from the nucleus. This leaves them with only a weak attraction to the protons at the center, and thus free to engage in bonding and other chemical activities. Without that minus sign introduced as electrons swap positions, there would be no stars, planets or life.

Electrons are fermions, yet in superconductors they form into BECs. So how does that happen? In 1956, Leon Cooper showed how electrons moving through a metal can bind together in pairs and acquire the characteristics of a boson. If all the electrons in a metal crystal form into such Cooper pairs, these bosons will come together to form, as in superfluid helium, one giant particle—a BEC.

The main consequence of this is a total lack of electrical resistance. In normal metals, resistance arises from electrons bumping into the metal ions bouncing around. But once a metal becomes a superconductor, the electron-pair condensate is in its lowest possible state. That means it cannot dissipate energy and, once the Cooper pairs are made to flow in an electrical current, they simply keep flowing. The only way to disturb superconductivity without raising the temperature is to add energy another way, for example by applying a sufficiently strong magnetic field.

Though superfluids and superconductors are bizarre enough, they are not the limit of the quantum world's weirdness, it seems. "There is yet another level of complexity," says Hinds. That complexity comes into play below 1 K and at more than 25 times Earth's atmospheric pressure, when helium becomes a solid. This form of helium plays havoc with our notions of solidity. Get the conditions right and you can make solids pass through each other like ghosts walking through walls.

Such an effect was first observed in 2004 by Moses Chan and Eunseong Kim at Penn State University in University Park, Pennsylvania. They set up solid helium in a vat that could rapidly rotate back and forth, inducing oscillations in the solid helium. They observed a resonant vibrational frequency which they interpreted as indicating that there were two solids in the vat, which were passing through each other.

Extreme superatoms

Superfluids, superconductors and supersolids owe their bizarre behavior to the formation of a sort of superatom inside them, known as a Bose–Einstein condensate (BEC).

But might it be possible to create such a state outside a liquid or solid? It took researchers many years, but in 1995 a team at the University of Colorado and the US National Institute of Standards and Technology, both based in Boulder, finally succeeded in coaxing a gas of rubidium into a BEC, its lowest possible quantum state. The breakthrough won team leaders Carl Wieman and Eric Cornell, together with Wolfgang Ketterle at the Massachusetts Institute of Technology, the 2001 Nobel prize for physics.

When Wieman and Cornell made their condensate, their lab briefly became home to the coldest place in the universe, just 20 nanokelvin above absolute zero. It wasn't the only BEC in the cosmos, though, even discounting superfluid or superconductor experiments that may have been taking place at exactly the same time.

In 2011, the Chandra X-ray telescope showed that the core of a neutron star called Cassiopeia A, which lies 11,000 light years away from Earth, is a superfluid. One teaspoon of neutron star material weighs six billion tons and the intense pressure from the outer layers is enough to squeeze the core into a BEC. Yet, despite the name, the core of a neutron star isn't exclusively made of neutrons; it contains a portion of protons too, which also form a BEC. You can think of this as a superfluid or, because the protons carry electrical charge, a superconductor.

Admittedly the two solids do not fit our usual definitions. One was made up of "vacancies," created when helium atoms shake free of the lattice that forms solid helium. The gaps left behind have all the properties of a real particle—they are so like real particles, in fact, that their quantum states can lock together to form a BEC. The solid helium is also a BEC, and it is these two condensates that pass through each other.

Chan and Kim's observation is still somewhat controversial; some researchers think there is a more prosaic explanation to do with deformations and defects in the helium lattice. "There is a lot of activity, several theory notions and experiments of interest, but no real agreement," says Robert Hallock of the University of Massachusetts at Amherst.

Nonetheless, even the fact that it might be possible to create solids that aren't really solid shows just how odd superstuff can get. And it's all because the world has a fundamental distinction at its heart. Everything, from human beings to weird low-temperature phenomena like liquids that defy gravity, stems from the fact that there are two kinds of particles: those that like to socialize, and those that don't. Sound familiar? Perhaps the quantum world isn't that different from us after all.

Pathways to cosmic oblivion

Crunch, snap, rip, fade . . . No, it's not a new breakfast cereal, but four ways our universe might end its days. What better way to conclude our exploration of nothing than to ask what will ultimately happen to all the matter that appeared in the big bang. Our guide to the finale is Stephen Battersby.

The future ain't what it used to be. Cosmologists were once confident they knew how the universe would end: it would just fade away. An ever colder, ever dimmer cosmos would slowly wind down until there were only cinders where the stars once shone. But that's history.

Today's science suggests many different possible futures. Cosmic cycles of death and rebirth might be on the cards, or a very peculiar end when the vacuum of space suddenly turns into something altogether different. The universe might collapse back in on itself in a big crunch. Or we could be in for an even more violent end called the big rip. Or a weird pixellation—the big snap. Or find our whole universe pouring down a wormhole (the big trip). The slow drift into darkness is still a contender, but fear not: that long night could be a lot more interesting than you might think—imagine the cosmos filled with giant diamonds.

Why this wealth of possibilities? Until recently, the dominant force in the universe seemed to be the gravity of stars and other matter, and that meant there were only two options. Either the universe was dense enough for gravity to halt the expansion from the big bang and pull everything back together in a big crunch or else it wasn't, in which case the expansion would carry on for ever.

Most cosmologists thought the latter possibility was the more likely. Then, in 1998, astronomers got a shock: they found that the universe's expansion isn't slowing down at all, but accelerating. Studies of the light of distant supernovae revealed that something stepped on the gas around 6 billion years ago. The finding seemed to seal our fate, condemning us to a headlong rush to oblivion. Acceleration means a universe that will become cold and boring much faster than we'd thought.

But researchers have now realized this gloomy outlook was premature, because no one knows what is causing the acceleration. Astronomers have named this mysterious force dark energy, but its origin and nature are a mystery. So how can anyone say what it is going to do in the future? "People started realizing that as long as we have no clue what dark energy is, we can't be so arrogant," says Max Tegmark of the Massachusetts Institute of Technology.

Although the long-term forecast is still open to debate, astronomers do at least agree on what will happen to our neighborhood in the near future. Things will become rather uncomfortable in about 6 billion years, when the sun swells to become a red giant, boiling the oceans away and possibly even swallowing Earth. Then our star will exhaust its nuclear fuel and shrink into a white dwarf roughly the size of Earth, leaving our old planet (if it still exists) cold enough to be covered in nitrogen ice. At least the view will be lovely: gas blown into space by the red giant will be energized by ultraviolet rays from the white dwarf, so for a while Earth will be surrounded by a glowing multicolored nebula.

Our galaxy is in for a rough time too. We are heading toward another, larger spiral called Andromeda, and we could collide in as little as 3 billion years. For a while, the merger will create a brilliant, elaborate hybrid galaxy, as streamers of stars are flung outward, and most of the loose gas in both galaxies is compressed to form bright new stars.

After only a couple of billion years more, those stellar tentacles will subside and the two delicate spirals will have merged into one great blob, an elliptical galaxy. Most of the free gas will have been used up, so relatively few new stars will form, except when small nearby galaxies in our local group are swallowed, each giving up its gas in a little puff of star formation.

Our collision with Andromeda will have a spectacular climax. At the center of the Milky Way is a giant black hole more than 3 million times the mass of the sun. Another in Andromeda is probably ten times the size. These two black holes will settle toward the center of the new galaxy, and there they will spiral together and eventually merge. The energy released will be tremendous, sending out a blast of light and X-rays, and a pulse of gravitational waves that will squeeze and stretch every star and planet.

Looking out into the universe we will see other galaxies moving away from our elliptical home, dragged ever faster by the hand of dark energy. But for how long? That depends on the nature of dark energy. For instance, its energy density could decrease with time. Some theorists have even come up with model universes where the dark energy becomes negative. As positive dark energy has repulsive gravity, negative dark energy would have attractive gravity, like ordinary matter.

If that happens in our universe, the consequences will be extreme. First acceleration will slow, and then dark energy will begin to really put the brakes on. Expansion will eventually halt, and then reverse, so that galaxies rush back toward each other and start colliding at ferocious speeds. Eventually, everything will be crushed together in a big crunch, unimaginably dense and hot, like the big bang in reverse.

That won't happen for a while, though. Those observations of distant supernovae, which trace the expansion of space over time, show that if dark energy is fading, it can't be doing so very fast. Andrei Linde of Stanford University in California has calculated that we are safe from a big crunch for at least 25 billion years, almost twice the age of the universe today.

But an even more grisly end could be in store. In 2003, Robert Caldwell of Dartmouth College in New Hampshire explored the opposite idea: that dark energy could become stronger. This exotic flavor of dark energy is called phantom energy. The expansion of space makes phantom energy increase, and phantom energy makes space expand even faster, setting up a devastating positive-feedback loop he calls the big rip.

If Caldwell is right, then a crisis could arrive in as little as 40 billion years from now. It would be perhaps the most watchable doomsday that cosmologists have imagined, not entirely unlike the spectacle laid on in Douglas Adams's *The Restaurant at the End of the Universe*.

Roughly 60 million years before the end, the phantom repulsion becomes strong enough to tear our galaxy apart. Then, just months before the end, the real show begins. Let's assume that by this point we have found ourselves a new home in a solar system not unlike this one. We will first see the outer planets fly away one by one. Next our adopted Earth will be torn from its sun. Less than an hour from the end, the sun will explode, and minutes later Earth will be ripped apart too. We might just be able to keep watching until a fraction of a second before the end, but presumably not long enough to see the destruction of molecules and atoms at around t-minus 10^{-19} seconds, when the phantom overpowers all electromagnetic forces. Neither will we see the subsequent shredding of nuclei, protons and neutrons. Pity.

Phantom energy could have a different outcome if our universe contains even a single wormhole. Wormholes are like tunnels in spacetime, possibly connecting one universe with another, and they would feed on phantom energy, and grow. A phantom-fed wormhole could grow large enough to swallow the whole universe, according to Pedro Gonzalez-Diaz at the Institute of Mathematics and Fundamental Physics, CSIC, Madrid. Gonzalez-Diaz calls this the big trip. It is not clear where the trip would take us.

But there doesn't have to be anything exotic about dark energy. The most conservative theory—what cosmologists call vanilla flavor—suggests that a given volume of vacuum has an inherent fixed energy, often called the cosmological constant. Many experts would bet that this kind of dark energy is what's causing the expansion to accelerate, and particle physicists even have a partial explanation for it: according to quantum mechanics, countless ephemeral subatomic particles are constantly popping in and out of existence, even in a vacuum, and their energy might add up to something. The only problem is that physicists struggle to explain the observed value of about 1 nanojoule per cubic meter. They can see how the particles' energy might cancel to zero or add up to a huge value, but not to next to nothing.

Nevertheless, this remains the most popular flavor of dark energy among cosmologists. If dark energy stays constant, our universe will steer carefully between crunch and rip.

Such a middle-of-the-road future may yet have a radical finale. According to quantum mechanics, the total amount of information in the universe should be constant. If space keeps expanding, that might make things uncomfortable, says Tegmark. The universe might eventually become pixellated, with information spread too thinly to support familar physics. Everything would disintegrate, in an event Tegmark has named the big snap. He has his doubts about this, though, suspecting that it only illustrates how we have no understanding of information at the most fundamental level.

If we avoid the big snap, then vanilla dark energy could lead to a long and lonely future. Acceleration will soon steal most of the universe away, as the increasing expansion of space carries other galaxies beyond our view. Their light will no longer reach us, because it is being dragged back over our cosmological horizon like a tortoise on a treadmill. According to Fred Adams of the University of Michigan in Ann Arbor, every other galaxy will have been pulled out of sight in a couple of hundred billion years.

Then we will be all alone, the observable universe reduced to our one elliptical galaxy, and a dingy one at that. There will be only a trace of free gas left to make new stars. Adams has calculated that even that will be used up after about a hundred trillion years, and all nuclear-powered stars will have gone out. A little faint infrared radiation will come from stars called brown dwarfs, which are too small to ignite fusion in their cores. Other stars will be reduced to dense, dead remnants—black holes, neutron stars and aging white dwarfs, slowly dimming to black. Our sun will become one of these black dwarfs: a single crystal of carbon, like an ultra-dense diamond, with a surface cool enough to touch.

Occasional flares will lift the gloom, when brown dwarfs collide to form a new star, or a black hole shreds a stellar carcass. Once in a trillion years, two relatively heavy black dwarfs will collide and explode as a supernova.

Every now and again, a star will be thrown out of the galaxy after a close encounter with another star. The whole galaxy will dissipate in about a hundred quintillion (10^{20}) years.

Now our observable universe is reduced to a diaspora of dead stars, loosely centered on a massive black hole surrounded by a cloud of dark matter. If there is a remnant of Earth, then for a while it might trail after the black dwarf that was once our sun. But the system will slowly lose angular momentum by emitting gravitational waves, and Earth's cinder will eventually spiral in to hit the sun's.

Meanwhile, dark energy will still be at work. Each star will see all its old companions disappear over the horizon one by one. Our black dwarf will be in a universe of its own.

After that, it gets a lot more speculative, but here's the best guess. Particle physicists suspect that protons are unstable and probably only last between 10^{33} and 10^{45} years. As protons decay into their constituent quarks, all the black dwarfs, neutron stars and planets will crumble away, leaving behind nothing but loose photons, neutrinos, electrons and positrons. Even black holes eventually evaporate, by a process called Hawking radiation, although that takes even longer—more than 10^{86} years for our central black hole.

And then? Dark energy continues working, even on these ashes. One day, every single particle will find itself alone inside its own horizon.

Aside from crunches, rips, trips, snaps and this lonely death, there is another possibility, a path almost parallel to that of the cosmological constant, but fractionally less bleak. In the cosmologists' models, some kinds of dark energy gradually fade in strength, but never become negative. Among them are defects in space-time that might be left over from the big bang, and a kind of energy field called quintessence. Just like the cosmological constant, these flavors would give us a chilly future where no stars shine and all solid bodies eventually dissipate into a cloud of fundamental particles. However, the acceleration will eventually tail off, so it won't isolate every particle within its own horizon. Particles could still interact, albeit at a glacial rate, and some kind of chilly life might just be able to cling to existence. Cold comfort, perhaps.

To find out which of these paths we will take, astronomers are examining the nature of dark energy. If they can pin down how space has expanded in the past, and learn what dark energy is really made of, we should have a clue to the future.

The favored oracles are distant stellar explosions known as type Ia supernovae. These supernovae are all of about the same power, so measuring both their apparent brightness and their distance tells us how much space has expanded since they went off. Astronomers are gathering more and more observations from the ground, and a proposed space telescope called WFIRST could spot thousands of type Ia supernovae.

To complement these observations, other astronomers are using galaxies to peer into dark energy's past. Because dark energy counteracts matter's tendency to clump together, its strength will have affected the number of galaxy clusters that formed at different stages of cosmic history. The best hope of tracing enough ancient clusters to pin down this history is an instrument called the Large Synoptic Survey Telescope, which could be running by 2021.

The results could tell us what the future holds. It may seem like a poor set of options—to be crushed, ripped apart or evaporated away. But in fact none of these scenarios need be the uttermost end. The universe, and perhaps even life, could survive any one of them.

In a big crunch, everything will be squashed into a super-hot, super-dense sea of radiation. It is certainly not going to be healthy for humans, but nobody knows what physics does when stuff gets that hot, so it's hard to predict what would happen to the universe itself. "Re-expansion is a possibility," says Linde. Since as far back as the 1930s, physicists have played with this idea, and if the universe can bounce, then maybe our own big bang was preceded by a crunch. It could happen again and again, big crunch leading to big bang and so on. A theory called loop quantum gravity actually predicts that a contracting space-time should bounce back.

It is an enticing idea, but there's a catch. Oscillating universes are vulnerable to a fatal disease: a plague of black holes. Holes survive the crunch, and in each cycle they grow. "They keep getting bigger and bigger till they swallow the whole universe," says Katherine Freese of the University of Michigan at Ann Arbor. But she may have a cure. With Matthew Brown, now at Lincoln Laboratory, Kwajalein, in the Marshall Islands, and William Kinney at the University of Buffalo, New York, Freese has concocted a new kind of oscillating universe that is immune to black hole disease.

Oddly, the prescription is a big rip. A dose of phantom dark energy tears everything apart—even black holes, effectively making them boil away. The cure may sound worse than the disease, but in fact this big rip can be repaired. This crazy-sounding idea is based on a respectable speculation, the "braneworld" model, in which our universe of three space and one time dimensions is like a membrane (or "brane") floating in higher-dimensional space.

In Freese's model, when phantom energy skyrockets to create the rip, it disturbs fields in the higher dimensions outside our "brane." They then transform the phantom energy, turning it negative and making our universe start to recollapse. Although all stars, planets and other structures from before the rip are gone, new objects might form during the collapse. If astronomers are among them, they will look back in time and discover a kind of big mend.

Then, as the universe reaches a big crunch, the energy density of ordinary matter and radiation soars. Fields in the higher dimension react again, making the contraction bounce back to become the expansion of a new big bang. All this may be rather contrived, but at least it shows there is a possibility that neither crunch nor rip need be the end of everything.

Freese suspects that her model universe would eventually run out of steam and stop bouncing. In contrast, the "ekpyrotic" universe (the name derives from the Greek word for conflagration) devised by Paul Steinhardt of Princeton University and Neil Turok, now at the Perimeter Institute in Waterloo, Canada, is supposed to be eternal. It's another braneworld model, with the twist that our brane is not the only one. Just a fraction of a millimeter away along the fifth dimension there is another universe. "They can collide from time to time like a pair of cymbals," says Turok. When they do their kinetic energy turns into a blast of radiation that we call a big bang.

Critics point out that when the branes collide, everything becomes infinitely dense, so the equations break down and the theory doesn't make much sense. Turok and his collaborators have now published a paper using M-theory—string theory's big brother—to show that this doesn't happen, but their idea remains controversial.

So what are your chances of surviving the cymbal-crashing big splat? Well, all the particles in any object would briefly become massless and fly apart at the speed of light, so you'd get rather scrambled, but it is possible that life could survive. "We would have to figure out how to preserve all our memories and information in the form of radiation," says Turok. "If you could imagine making a computer out of light, you could transmit it through the big bang and recover it on the other side."

There might be a similar escape route through Freese's rips and crunches—at a fundamental level information is preserved, so conceivably there is a way to encode ourselves into the next cycle of creation. Even in the long, slow decline of a constant dark energy there's a chance the universe could reinvent itself (see box "Quantum resurrection"). Who says you can't live for ever?

It's time for a confession. All of these forecasts are only local: they apply to the bit of our universe that lies within our cosmological horizon. But it is quite possible that the universe is truly infinite. Far beyond the horizon, conditions may be very different. Even the constants of physics may be different, and perhaps some of those regions may be more durable. In some models an infinite cosmos is constantly spawning new big bangs.

None of this can affect us, or have any bearing on our future—unless, perhaps, we somehow learn to manipulate wormholes in space-time and tunnel to freedom, moving to a fresh region of the cosmos whenever the old one gets tired. But even if we can't escape our local universe, at least it might be reassuring to think that the cosmos itself is immortal.

Quantum resurrection

Even if we face a future in which the cosmological constant reduces us all to a set of isolated particles, there is some hope. Quantum mechanics tells us that there are always fluctuations in any system. Energy fields waver at random, and particles can appear out of the vacuum. Large fluctuations are very rare, and you'd have to wait an extraordinary length of time for something big to appear—a whole atom or molecule, say.

But if our future is infinite, time is not a problem. Eventually, anything could spontaneously pop into existence. Most of these things will be senseless messes, but a vanishingly small proportion will be people, planets, galaxies, and five-mile-long models of your left arm made from gold. "In an infinite amount of time, I will reappear. A crazy thought, but true," says Katherine Freese of the University of Michigan at Ann Arbor.

How about a whole new universe? Sean Carroll, now at the California Institute of Technology in Pasadena, thinks that random fluctuations could spark a new big bang. He's even worked out how long we might have to wait for it, something in the region of $10^{10^{56}}$ years, or a 1 followed by 10^{56} zeros.

This dwarfs all the timescales we have met so far—it's even impossible to write down in conventional longhand notation. It is hard to imagine how any kind of life could survive long enough to take advantage of the new universe. Unless, perhaps, it can find a technology that will trigger the new big bang, restarting the cycle of cosmic life and death.

Acknowledgments

The idea of a collection of essays about nothing goes back to 1998. The deputy features editor of *New Scientist* at the time, Gabrielle Walker, strolled into the office after yet another trip, this one to San Francisco, and told us of a museum exhibition she'd seen on nothing. "What can we do with that?" she asked. We leapt on the idea, creating a best-selling issue of the magazine. So the biggest thank you must go to Gabrielle for keeping her eyes open and her wits about her. (Only two articles from that original collection made it through to this book, though: science moves quickly these days.)

At the end of 2010, we thought it would be a good idea to look at nothing once more. The entire London staff of *New Scientist* turned out for a brainstorm to create ideas that would top the first batch. It was a fantastic success, creating another special issue and ultimately this book. So thanks to everyone who offered suggestions and added to *New Scientist*'s rich and unselfish culture of ideas.

I must thank all the authors for their inspirational work, and the editors who guided them to create great stories. More than a dozen editors worked on the essays in this book, all of them talented people. A few are conspicuous for being both authors and editors here—Stephen Battersby, Michael Brooks and Richard Webb.

Special thanks go to Valerie Jamieson for pulling many of these articles together, for her advice and for being more talented than she realizes. Thanks also to Paul Marks for proposing "The hole story," which he never got to write.

Editors can take articles so far, but it is the subeditors who hone them to perfection. Here again many "subs" have contributed. Thanks to every one of them, and especially to Liz Else, who has an eye for the intriguing and the knack of bringing it out, John Liebmann, for setting such high standards (sadly John died while this book was being prepared), and Chris Simms, who put me right while subbing a couple of stories that had not appeared in the magazine.

Finally, at *New Scientist* I want to thank Nigel Hawkin and Dave Johnston for creating the original artwork for the illustrations (all redrawn for the book)—each one is worth more than a thousand words. And thanks too to Ellie Harris for her eagle eyes at the proofreading stage.

I want to say a big thank you to Andrew Franklin at Profile. I dropped the idea of a book about nothing as an afterthought while leaving his office. He cogitated for a few days and came back with a big "Yes!" You wouldn't be reading this without his enthusiasm. Final thanks to Paul Forty, whose good grace and humor are beyond price.

These are exceptional stories—great science, written well—for which I can claim no credit. That belongs to the writers and editors. I am, however, responsible for any errors created in updating the essays. The fault for these is mine alone.

Jeremy Webb

About the contributors

Publication dates are given for essays that have previously appeared in New Scientist.

Philip Ball ("Impossible reaction," published January 21, 2012) is a freelance writer. He previously worked for over twenty years as an editor for *Nature*. He has written many books on the interactions of the sciences, arts and wider culture, including *The Self-Made Tapestry: Pattern Formation in Nature, H_2O: A Biography of Water, Critical Mass* and *The Music Instinct*.

Stephen Battersby ("Pathways to cosmic oblivion," published February 5, 2005) is a freelance science writer, quiz-question setter and consultant for *New Scientist*. He covers most sides of science, but has a soft spot for icy moons.

Michael Brooks ("Placebo power," published August 20, 2008, and "The world of superstuff," published January 14, 2012) is a *New Scientist* consultant and the author of *13 Things That Don't Make Sense, The Secret Anarchy of Science* and *Can We Travel Through Time?* He holds a PhD in quantum physics and is a regular contributor to a variety of newspapers and magazines.

Marcus Chown ("The big bang," published October 22, 1987) is an award-winning writer and broadcaster. His books include *Quantum Theory Cannot Hurt You, Solar System for iPad* and *Tweeting the Universe*. His latest book is *What A Wonderful World: One man's attempt to explain the big stuff*.

Andy Coghlan ("The workout pill," published August 25, 2012) has been reporting breakthroughs in science and technology for *New Scientist* since 1986, focusing mainly on biomedical news. His award include prizes presented by the UK Medical Journalists' Association, the Association of British Science Writers and the American Society for Microbiology. He still runs up and down the stairs daily, and has so far avoided the need to take statins.

Paul Davies ("The day time began," published April 27, 1996, and "The turbulent life of empty space," published November 19, 2011) is director of the Beyond Center for Fundamental Concepts in Science at Arizona State University in Tempe. His latest book is *The Eerie Silence: Are we alone in the universe?*

Michael de Podesta ("Absolute zero," published June 22, 2013) is a physicist and temperature expert at the National Physical Laboratory in London.

Per Eklund ("Out of thin air") is associate professor of materials physics at Linköping University in Sweden. He is one of the editors of the journal *Vacuum* and has developed and teaches courses on vacuum science and technology. He is an elected member of the Young Academy of Sweden.

David E. Fisher ("Putting the idle to work," published November 19, 2011) is professor emeritus of geological science and cosmochemistry at the University of Miami and the author of *Much Ado About (Practically) Nothing: A history of the noble gases*.

Douglas Fox ("The secret life of the brain," published November 5, 2008) is a science journalist based in California. He has written for *New Scientist, Discover, Popular Mechanics, Scientific American, Esquire, National Geographic* and *The Christian Science Monitor*. His stories have garnered national award from the American Association for the Advancement of Science and the American

Society of Journalists and Authors, and have been anthologized in *The Best American Science and Nature Writing*.

Linda Geddes ("Banishing consciousness," published November 29, 2011) is an award-winning British journalist and author who writes about the science of sex, death and everything in between. She graduated from the University of Liverpool with a first class degree in cell biology and is a reporter for *New Scientist*. Her book *Bumpology: The myth-busting pregnancy book for curious parents-to-be* was published in January 2013.

David Harris ("Vacuum packed," published February 18, 2012) is a science writer and editor based in Palo Alto, California.

Nigel Henbest ("Into the void," published April 25, 1998) is an award-winning author and television producer, specialising in astronomy and space. He researched in radio astronomy at the University of Cambridge, and has since penned 40 books and over 1,000 articles, with translations into 27 languages. His 60-plus TV programmes have been screened worldwide.

Valerie Jamieson ("Boring-ology," published December 24, 2005) is features editor at *New Scientist* and used to smash atoms at the DESY laboratory in Hamburg.

Jonathan Knight ("Busy doing nothing," published April 25, 1998) traded in a successful career as a journalist to teach biology and writing at San Francisco State University, California.

Richard A. Lovett ("Get up, get out of bed," published August 20, 2005) has a long interest in sports and fitness. He has cycled solo across North America and skied 160 kilometers north of the Arctic Circle in Greenland. Working through Team Red Lizard running club, in Portland, Oregon, he has coached two women to the US Olympic marathon trials. He also writes science fiction, sometimes on sports themes.

Jo Marchant ("Heal thyself," published August 30, 2011) is a freelance science journalist based in the UK. She writes on topics from the future of genetic engineering to underwater archaeology and is the author of two books, *Decoding the Heavens* and *The Shadow King*. Her third book, *Heal Thyself*, is due to be published in 2014.

Helen Pilcher ("When mind attacks body," published May 13, 2009) is a freelance science writer and performer based in the UK. She writes serious and quirky science for the likes of *New Scientist* and *Nature*, and funny stuff for comedy shows. She has a PhD in neuroscience.

Laura Spinney ("Wastes of space," published May 14, 2008) is a science writer and novelist. Her latest book, *Rue Centrale*, is a factual portrait in French of the city of Lausanne.

Ian Stewart ("Zero, zip, zilch," published April 25, 1998, "Ride the celestial subway," published March 27, 2006, and "Nothing in common," published November 19, 2011) is an emeritus professor of mathematics at Warwick University and a Fellow of the Royal Society. He was awarded the Society's Faraday Medal in 1995. His recent books include *17 Equations That Changed the World*, *The Great Mathematical Problems* and *The Science of Discworld IV* (with Terry Pratchett and Jack Cohen).

Jeremy Webb is editor-in-chief of *New Scientist*, where he has worked in various roles including ten years as editor. Before joining *New Scientist*, he wrote and edited news for television and *Pulse*, a UK newspaper for family doctors. His media career began in the BBC, where he worked as a sound engineer and producer of radio science programmes. This varied CV means he knows a little about a lot of scientific fields, but a lot about—well, nothing.

Richard Webb ("From zero to hero" and "The hole story," both published November 19, 2011) is deputy features editor at *New Scientist*. Before that he was an editor at *Nature*, and before that a particle physicist at CERN, where he studied the very lively nothingness that is the interior of the proton.

Notes

Secret life of the brain

1. "A default mode of brain function," *Proceedings of the National Academy of Sciences*, 98, pp. 676–82, doi: 10.1073/pnas.98.2.676.
2. "Wandering minds: the default network and stimulus-independent thought," *Science*, 19 January 2007, vol. 315, no. 5810, pp. 393–5, doi: 10.1126/science.1131295.
3. "Intrinsic functional architecture in the anaesthetized monkey brain," *Nature*, May 3, 2007, 447, pp. 83–6, doi: 10.1038/nature05758.
4. "Persistent default-mode network connectivity during light sedation," *Human Brain Mapping*, July 2008, vol. 29, issue 7, pp. 839–47, doi: 10.1002/hbm.20537.

   Horovitz, S., et al., "Low frequency BOLD fluctuations during resting wakefulness and light sleep: A simultaneous EEG-fMRI study," *Human Brain Mapping*, 2008, 29, pp. 671–82, doi: 10.1002/hbm.20428.
5. Pagnoni, G., et al., "Thinking about not-thinking: neural correlates of conceptual processing during Zen meditation." *PLoS ONE*, 2008, 3(9): e3083, doi: 10.1371/journal.pone.0003083.

Heal thyself

1. Kaptchuk, Ted J., et al., "Placebos without deception: a randomized controlled trial in irritable bowel syndrome,"

PLoS ONE, 2010, 5(12): e15591, doi: 10.1371/journal.pone.0015591.

2. Aspinwall, Lisa G., et al., "The value of positive psychology for health psychology: progress and pitfalls in examining the relation of positive phenomena to health," *Annals of Behavioral Medicine*, February 2010, 39(1), pp. 4–15.
3. Chida, Y., and A. Steptoe, "Positive psychological well-being and mortality: a quantitative review of prospective observational studies," *Psychosomatic Medicine*, September 2008, 70(7), pp. 741–56, doi: 10.1097.
4. Taylor, Shelley E., et al., "Are self-enhancing cognitions associated with healthy or unhealthy biological profiles?," *Journal of Personality and Social Psychology*, October 2003, 85(4), pp. 605–15, doi: 10.1037/0022–3514.85.4.605.
5. Sherman, D. K., et al., "Psychological vulnerability and stress: the effects of self-affirmation on sympathetic nervous system responses to naturalistic stressors," *Health Psychology*, September 2009, 28(5), pp. 554–62, doi: 10.1037/a0014663.
6. Hawkley, Louise C., "Loneliness matters: a theoretical and empirical review of consequences and mechanisms," *Annals of Behavioral Medicine*, 2010, 40(2), pp. 218–27, doi: 10.1007/s12160–010–9210–8.
7. Jacobs, Tonya L., et al., "Intensive meditation training, immune cell telomerase activity, and psychological mediators," *Psychoneuroendocrinology*, June 2011, 36(5), pp. 664–81.
8. Hölzel, Britta K., et al., "Stress reduction correlates with structural changes in the amygdala," *Social, Cognitive and Affective Neuroscience*, 2010, 5(1), pp. 11–17, doi: 10.1093/scan/nsp034.
9. Whorwell, Peter J., "Hypnotherapy for irritable bowel syndrome: The response of colonic and noncolonic

symptoms," *Journal of Psychosomatic Research*, 2008, 64(6), pp. 621–3, doi: 10.1016/j.jpsychores.2008.02.022.

10. Mendoza, M. Elena, et al., "Efficacy of clinical hypnosis: A summary of its empirical evidence," *Papeles des Psicólogon*, 2009, 30, p. 98.

11. Lissoni, P., et al., "A spiritual approach in the treatment of cancer: relation between faith score and response to chemotherapy in advanced non-small cell lung cancer patients, *in vivo*," 2008, 22, pp. 577–82.

12. Chida, Y., et al., "Religiosity/spirituality and mortality, psychotherapy and psychosomatics," 2009, 78, pp. 81–90, doi: 10.1159/000190791.

13. Kohls, Nikola, et al., "Spirituality: an overlooked predictor of placebo effects?," *Philosophical Transactions of The Royal Society B*, 2011, 366(1572), pp. 1838–48.

Placebo power

1. Benedetti, F., et al., "Open versus hidden medical treatments: the patient's knowledge about a therapy affects the therapy outcome," *Prevention & Treatment*, 2003, 6, p. 1a, doi: 10.1037/1522–3736.6.1.61a.

2. Kaptchuk. Ted J., et al., "Components of placebo effect: randomised controlled trial in patients with irritable bowel syndrome," *BMJ*, 2008, 336, pp. 999–1003, doi: 10.1136/bmj.39524.439618.25.

3. Colloca, Luana, et al., "Learning potentiates neurophysiological and behavioral placebo analgesic responses," *Pain*, 2008, 139, pp. 306–14, doi: 10.1016/j.pain.2008.04.021.

4. Miller, Franklin G., and Ted J. Kaptchuk, "The power of context: reconceptualising the placebo effect," *Journal of the Royal Society of Medicine*, 2008, 101, pp. 222–5, doi: 10.1258/jrsm.2008.070466.

5. Hróbjartsson, Asbjørn, and Peter C. Gøtzsche, "Placebo interventions for all clinical conditions," The Cochrane Library, doi: 10.1002/14651858.CD003974.pub3.

Wastes of space

1. Bollinger, R. Randal, et al., "Biofilms in the large bowel suggest an apparent function of the human vermiform appendix," *Journal of Theoretical Biology*, December 21, 2007, 249(4), pp. 826–31, doi: 10.1016/j.jtbi.2007.08.032.
2. Nesse, Randolph M., and George C. Williams, "Evolution and the origins of disease," *Scientific American*, November 1998, pp. 86–93.
3. Lucas, Peter W., "Facial dwarfing and dental crowding in relation to diet," International Congress Series, June 2006, 1296, pp. 74–82, doi: 10.1016/j.ics.2006.03.041.

Banishing consciousness

1. Baars, Bernard J., "In the theater of consciousness: global workspace theory, a rigorous scientific theory of consciousness," *Journal of Consciousness Studies*, 1997, 4(4), pp. 292–309.
2. Alkire, Michael T., et al., "Consciousness and anesthesia," *Science*, 2008, 322(5903), pp. 876–80, doi: 10.1126/science.1149213.
3. Noirhomme, Quentin, et al., "Brain connectivity in pathological and pharmacological coma," *Frontiers in Systems Neuroscience*, December 20, 2010, doi: 10.3389/fnsys.2010.00160.
4. Ku, Seung-Woo, et al., "Preferential inhibition of frontal-to-parietal feedback connectivity is a neurophysiologic correlate of general anesthesia in surgical patients," *PLoS One*, doi: 10.1371/journal.pone.0025155.

5. Davis, Matthew H., et al., "Dissociating speech perception and comprehension at reduced levels of awareness," *Proceedings of the National Academy of Sciences*, 104, pp. 16032–7, doi: 10.1073/pnas.0701309104.

The hole story

1. http://www.aip.org/history/ohilist/4817.html
2. http://www.pbs.org/transistor/album1/shockley/

When mind attacks body

1. Meador, C. K., "Hex death: voodoo magic or persuasion?," *Southern Medical Journal*, March 1992, 85(3), pp. 244–7.
2. Barsky, A. J., et al., "Nonspecific medication side effects and the nocebo phenomenon," *Journal of the American Medical Association*, February 6, 2002, 287(5), pp. 622–7.
3. http://www.newscientist.com/article/dn16743.
4. Lorber, W., et al., "Illness by suggestion: expectancy, modeling, and gender in the production of psychosomatic symptoms," *Annals of Behavioral Medicine*, February 2007, 331, pp. 112–16.
5. Scott, D. J., et al., "Placebo and nocebo effects are defined by opposite opioid and dopaminergic responses," *Archives of General Psychiatry*, February 2008, 65(2), pp. 220–31, doi: 10.1001/archgenpsychiatry.2007.34.
6. Eaker, E. D., et al., "Myocardial infarction and coronary death among women: psychosocial predictors from a 20-year follow-up of women in the Framingham Study," *American journal of epidemiology*, April 15, 1992, 135(8), pp. 854–64.
7. Kharabsheh, Saad, et al., "Mass psychogenic illness following tetanusdiphtheria toxoid vaccination in Jordan," *Bulletin of the World Health Organization*, 2001, 79, pp.

764–70, http://www.who.int/bulletin/archives/79(8)764.pdf.

Ride the celestial subway

1. Koon, Wang Sang, et al., "Constructing a low energy transfer between Jovian moons," *Contemporary Mathematics*, 2002, 292, pp. 129–45.
2. Dellnitz, M., et al., "On target for Venus–set oriented computation of energy efficient low thrust trajectories," *Celestial Mechanics and Dynamical Astronomy*, 2006, 95(1–4), pp. 357–70.

Vacuum packed

1. Wilson, C. M., et al., "Observation of the dynamical Casimir effect in a superconducting circuit," *Nature*, 2011, 479, pp. 376–9, doi: 10.1038/nature10561.
2. Lamoreau, S. K., "Demonstration of the Casimir force in the 0.6 to 6*µm* range," *Physical Review Letters*, 1997, 78, pp. 5–8, doi: 10.1103/PhysRevLett.78.5.
3. Sushkov, A. O., et al., "Observation of the thermal Casimir force," *Nature Physics*, 2011, 7, pp. 230–33, doi: 10.1038/nphys1909.
4. Maslovski, Stanislav I., and Mário G. Silveirinha, "Mimicking Boyer's Casimir repulsion with a nanowire material," *Physical Review A*, 2011, 83, p. 022508, doi: 10.1103/PhysRevA.83.022508.
5. Munday, J. N., et al., "Measured long-range repulsive Casimir–Lifshitz forces," *Nature*, 2009, 457, pp. 170–73, doi: 10.1038/nature07610.
6. Rugh, Svend, and Henrik Zinkernagel, "The quantum vacuum and the cosmological constant problem," arXiv:hep-th/0012253, http://philsci-archive.pitt.edu/id/eprint/398.

7. Moore, G. T., "Quantum theory of the electromagnetic field in a variable-length one-dimensional cavity," *Journal Mathematical Physics*, 1970, 11, 2679, doi: 10.1063/1.1665432.

8. Wilson, op. cit.; see note 1.

Boring-ology

1. Bean, W. B., "Nail growth: Thirty-five years of observation," *Archives of Internal Medicine*, 1980, 140, pp. 73–6.

Putting the idle to work

1. Khriachtchev, Leonid, et al., "A stable argon compound," *Nature*, 2000, 406, pp. 874–6, doi: 10.1038/35022551.

2. Seidel, Stefan, and Konrad Seppelt, "Xenon as a complex ligand," *Science*, 2000, 290, pp. 117–18, doi: 10.1126/science.290.5489.117.

The workout pill

1. Blair, S. N., "Physical inactivity: the biggest public health problem of the 21st century," *British Journal of Sports Medicine*, 2009, 43, pp. 1–2.

2. *Diabetes in the UK 2004*, http://www.diabetes.org.uk/Professionals/Publications-reports-and-resources/Reports-statistics-and-case-studies/Reports/Diabetes_in_the_UK_2004/.

3. Flegal, K. M., et al., "Prevalence of obesity and trends in the distribution of body mass index among US adults, 1999–2010," *The Journal of the American Medical Association*, 2012, 307(5), pp. 491–7, doi: 10.1001/jama.2012.39.

4. Al-Shayji, I. A. R., et al., "Effects of moderate exercise on VLDL1 and Intralipid kinetics in overweight/obese middle-aged men," *American Journal of Physiology and Metabolism*, 2012, 302(3), pp. E349–55, doi: 10.1152/ajpendo.00498.2011.

5. Diabetes Prevention Program Research Group, "Reduction in the incidence of type 2 diabetes with lifestyle intervention or Metformin," *New England Journal of Medicine*, 2002, 346, pp. 393–403, doi: 10.1056/NEJMoa012512.

6. Richter, E. A., et al., "Muscle glucose metabolism following exercise in the rat: increased sensitivity to insulin," *The Journal of Clinical Investigation*, 1982, 69(4), pp. 785–93, doi: 10.1172/JCI110517.

7. Wojtaszewski, J. F., et al., "Insulin signaling and insulin sensitivity after exercise in human skeletal muscle," *Diabetes*, 2000, 49(3), pp. 325–31, doi: 10.2337/diabetes.49.3.325.

8. Funai, K., et al., "Increased AS160 phosphorylation, but not TBC1D1 phosphorylation, with increased postexercise insulin sensitivity in rat skeletal muscle," *American Journal of Physiology, Endocrinology and Metabolism*, 2009, 297(1), pp. E242–51, doi: 10.1152/ajpendo.00194.2009.

9. Holmes, M. D., "Physical activity and survival after breast cancer diagnosis," *The Journal of the American Medical Association*, 2005, 293(20), pp. 2479–86, doi: 10.1001/jama.293.20.2479.

10. Slattery, M. L., and J. D. Potter, "Physical activity and colon cancer: confounding or interaction?," *Medicine and Science of Sports Exercise*, 2002, 34(6), pp. 913–19.

11. McTiernan, A., et al., "Effect of a 12-month exercise intervention on patterns of cellular proliferation in colonic crypts: a randomized controlled trial," *Cancer Epidemiology, Biomarkers & Prevention*, 2006, 15, p. 1588, doi: 10.1158/1055–9965.EPI-06–0223.

12. He, C., et al., "Exercise-induced BCL2-regulated autophagy is required for muscle glucose homeostasis," *Nature*, 2012, 481, pp. 511–15, doi: 10.1038/nature10758.

13. van Praag, H., et al., "Running increases cell proliferation and neurogenesis in the adult mouse dentate gyrus," *Nature Neuroscience*, 1999, 2, pp. 266–70, doi: 10.1038/6368.

14. Erickson, K. I., et al., "Exercise training increases size of hippocampus and improves memory," *Proceedings of the National Academy of Sciences*, 2011, 108(7), pp. 3017–22, doi: 10.1073/pnas.1015950108.

15. Chaddock, L., et al., "A neuroimaging investigation of the association between aerobic fitness, hippocampal volume, and memory performance in preadolescent children," *Brain Research*, 2010, 1358, pp. 172–83.

16. Steeves, J. A., "Energy cost of stepping in place while watching television commercials," *Medicine & Science in Sports & Exercise*, 2012, 44, pp. 330–35, doi: 10.1249/MSS.0b013e31822d797e.

17. Sui, X., et al., "Cardiorespiratory fitness and adiposity as mortality predictors in older adults," *The Journal of the American Medical Association*, 2007, 298(21), pp. 2507–16, doi: 10.1001/jama.298.21.2507.

18. Wen, C. P., and Xifeng Wu, "Stressing harms of physical inactivity to promote exercise," *The Lancet*, 2012, 380(9838), pp. 192–3, doi: org/10.1016/S0140-6736(12)60954-4.

Index

Numbers in *italics* indicate Figures.